HUMAN BIOLOGY

HIGHER GRADE COURSE NOTES

ANDREW MORTON

Leckie & Leckie

Published by Leckie & Leckie
8 Whitehill Terrace
St Andrews
Scotland, UK
KY16 8RN
Tel: 01334 475656
Fax: 01334 477392
E-mail: s.leckie@leckie-and-leckie.co.uk
Website: www.leckie-and-leckie.co.uk

Edited by   Alison Rogers

Special thanks to Julie Barclay, Alison Irving, Susanna Kirk, Frances McDonald, Bruce Ryan, Hamish Sanderson and Norman Stewart

ISBN 1-898890-11-0

A CIP Catalogue record for this book is available from the British Library.

Printed in Scotland by Inglis Allen.

Please note: Words in **bold type** are likely to be examined.

UNIT ONE
CELL FUNCTION AND INHERITANCE

This forty-hour Unit contains information on cell structure, cell metabolism, the activity of enzymes, chemical (tissue) respiration, cell transport, cell defence systems, immunity, DNA replication, protein synthesis, meiosis and genetics.

The Cell

The human body is made of many millions of cells, such as nerve cells, blood cells, muscle cells, bone cells and sex cells, each type having a specific function. All cells contain a fluid **cytoplasm** surrounded by a membrane called the **plasma membrane**. In the cytoplasm are many tiny **organelles** and a vast number of different chemical compounds. The chemical reactions which take place in the **organelles** and in the cytoplasm are referred to as the cell's **metabolism** and many of these metabolic reactions are **catalysed** (speeded up) by enzymes. A series of such reactions is called a **metabolic pathway**.

A typical cell (as seen under an electron microscope)

This diagram is magnified around 2000x. The cell is invisible to the naked eye and measures around 50 micrometres. One micrometre (1 μm) is a thousandth of a millimetre or 10^{-6}m.

Golgi body

Golgi vesicle

lysosome

centriole

ribosomes

rough endoplasmic reticulum (E.R.)

plasma membrane

nucleus

nucleolus

nuclear membrane (with pores)

mitochondrion

smooth endoplasmic reticulum (E.R.)

Each of the organelles has a specific function, as outlined in the table below:

Organelle	Brief outline of what it does
nucleus	contains DNA which acts as blueprint for the manufacture of proteins
nucleolus	involved in the synthesis of ribosomes
mitochondrion	site of production of energy-rich ATP by aerobic respiration
Golgi body	packaging of complex molecules for excretion by the cell
Golgi vesicle	small sac carrying complex molecules to plasma membrane
rough E.R.	sheets of membrane covered in ribosomes and involved in the synthesis and transport of proteins
smooth E.R.	sheets of membrane with no ribosomes; involved in the synthesis and transport of lipids
ribosomes	found on the E.R. and free in the cytoplasm, they are the site of protein synthesis
centriole	two cylindrical structures involved in the formation of the spindle fibres during mitosis and meiosis
lysosome	a sac derived from Golgi vesicles which contains digestive enzymes

Enzymes

A synthesis reaction

Enzymes are **proteins** manufactured by the ribosomes. Some enzymes are exported from the cell to act in, for example, the gut or the tears. Others are kept within the cell to catalyse the cell's metabolism. Without enzymes, the body's metabolism would be too slow at body temperature (37°C).

Enzymes act by combining with the **substrate** molecule(s) in a **highly specific** way so that one enzyme can normally only combine with one set of substrates. The substrates may be broken down (catabolism) or combined with others to make more complex molecules (anabolism). The substrates fit almost perfectly into an **active site** on the enzyme molecule rather in the same way as a key fits in a lock. When the reaction has taken place the fit is not quite so good and the product leaves the active site free for another reaction to take place. Some enzymes can **synthesise** or break down many hundreds of molecules per second, but their activity relies on the shape of the active site being maintained. If the temperature or pH of the surroundings vary from the **optimum** (ideal), then the enzyme molecule becomes **denatured** (altered) and no longer functions so effectively.

Graphs to show the effect of changes in pH, temperature, substrate concentration and enzyme concentration on the rates of enzyme reactions.

Action of Inhibitors

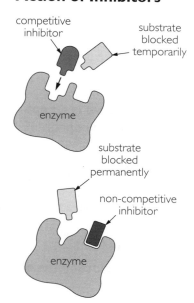

Some compounds, which have shapes similar to enzyme substrates, compete with substrates for the active sites on the enzyme molecules. Such substances tend to slow down the activity of the enzyme and are called competitive inhibitors. Other types of inhibitor become attached to other parts of the enzyme molecule and alter the whole structure of the enzyme permanently so that its active sites no longer function. Poisons such as cyanide, mercury and lead work in this fashion and are called non-competitive inhibitors.

Some enzymes do not function properly unless **activators** or **coenzymes** are present. Many **vitamins** and **minerals** act as activators; for this reason they are essential in the diet. In the gut some enzymes are secreted in an inactive form because they would otherwise digest the cells which produce them. Such enzymes are activated by other enzymes in the digestive juices.

Sometimes an enzyme is absent from a cell due to a **mutation** (a fault in the genes). In such cases the metabolic pathway is blocked and very often the effect is lethal. For example, the absence of the enzyme which converts the amino acid phenylalanine to the amino acid tyrosine results in a condition called **phenylketonuria** (**PKU**). The metabolic pathway is shown below.

The metabolic pathway for the production of melanin

Without *enzyme A*, phenylalanine accumulates in the blood and some of it is converted into phenylpyruvic acid which is excreted in the urine. The excess of both compounds disrupts the normal development of various organs, including the brain. Fortunately, there is now a blood test which can be carried out at birth to check for this condition. Testing babies shortly after birth for genetic defects such as PKU is referred to as **post-natal screening**. Children suffering from PKU are given a low phenylalanine diet for a number of years until their brains are fully developed. Sufferers of PKU also have lighter hair and lighter skins because they can only make melanin from tyrosine and not from phenylalanine. Tyrosine is obtained in food, as are all the other amino acids.

Proteins

Proteins are **organic compounds** composed of **hydrogen, oxygen, nitrogen** and **carbon** atoms with the addition of various other elements such as sulphur. They are complex molecules with three-dimensional structures which are unique for each protein. There are tens of thousands of different types of protein; each protein is composed of a large number of building blocks called **amino acids,** of which about twenty types are found in nature. These amino acids are polymerised (linked together) by **peptide bonds** to form chains called **peptides** or **polypeptides**. These in turn are linked by bonds such as **hydrogen bonds,** to form **secondary** and **tertiary** foldings which are important in the functioning of proteins.

Proteins have a multitude of roles to perform in the body, so much so that it would not be far from the truth to say we are 'made of protein'.

- All enzymes are proteins.
- Many hormones are proteins. For example: ADH (antidiuretic hormone) is a peptide; growth hormone is a protein; TSH (thyroid-stimulating hormone) is a glycoprotein; prolactin is a protein.
- Muscle contraction is brought about by the sliding action of two types of protein filament: **actin** and **myosin**. (See diagram below.)
- Haemoglobin is a protein which transports oxygen in the blood. Transferrin is a protein which carries iron.
- Antibodies are proteins which protect the body from foreign antigens.
- Fibrinogen and prothrombin are proteins which protect the body by clotting the blood at wounds.
- All membranes have a protein component.
- The cell has a cytoskeleton which is composed of protein. For example, microvilli, spindle fibres and cilia all maintain their shape because of their protein 'skeletons'.
- Bones, tendons, ligaments, skin and hair are all principally composed of proteins such as collagen, elastin and keratin.

Muscle Structure

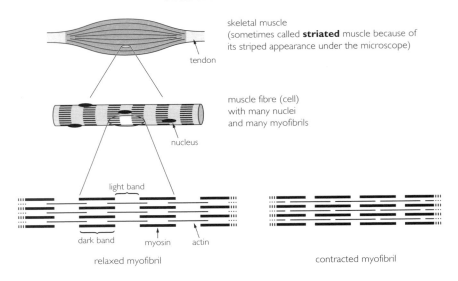

skeletal muscle
(sometimes called **striated** muscle because of its striped appearance under the microscope)

tendon

muscle fibre (cell)
with many nuclei
and many myofibrils

nucleus

light band

dark band   myosin   actin

relaxed myofibril   contracted myofibril

DNA, RNA and Protein Synthesis

Chromosome structure

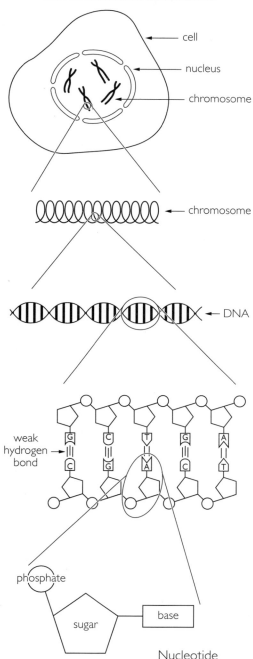

cell

nucleus

chromosome

chromosome

DNA

weak hydrogen bond

phosphate

sugar

base

Nucleotide

Deoxyribonucleic acid (DNA) is the chemical which carries the genetic code ('blueprint') for building almost all living things, including humans. When cells are dividing, DNA can be seen under light microscopes as immensely coiled threads called **chromosomes** which are found in the nucleus. DNA is not a protein itself but it dictates which proteins are made in our cells. The DNA of our sex cells is particularly important because it carries the genetic instructions from one generation to the next. DNA is a relatively simple molecule made up of millions of **nucleotides** arranged in two spiralling rows rather like a twisted ladder.

The spiral shape of the molecule was discovered by Watson and Crick in 1952 and is described as a **double helix**. Each nucleotide is composed of a **phosphate** molecule, a **sugar** molecule and a single **base** of which there are four types. The four bases exist in pairs and make up the 'rungs' of the double helix. **Adenine** always pairs with **thymine**; **guanine** always pairs with **cytosine**. The sequence of these bases dictates the manufacture of proteins in the cell.

To start the manufacture of protein, the DNA code is **transcribed** (copied) onto a strand of **messenger ribonucleic acid** (**mRNA**). mRNA carries the code out through the pores in the nucleus to the **ribosomes** in the cytoplasm where proteins are made. Ribosomes are found free in the cell or attached to sheets of membranes called rough **endoplasmic reticulum** (**E.R.**). Free ribosomes synthesise proteins for use within the cell. Attached ribosomes synthesise proteins for the cell membrane and for export.

RNA is similar to DNA but has a different base (**uracil** instead of thymine) and a different sugar (**ribose** instead of **deoxyribose**). Moreover, RNA exists as a **single strand** rather than a **double strand**. At the ribosomes, a second type of RNA called **transfer RNA** (**tRNA**) brings in amino acids for assembly into protein. tRNA exists in strands of which three particular bases, called **anti-codons**, correspond to the appropriate triplet **codons** in the mRNA.

The code is in triplets to allow for the coding of the twenty different types of **amino acid**. Since four bases can be arranged in threes in sixty-four different ways, many amino acids have more than one code, e.g. UUU and UUC both code for the amino acid phenylalanine. In addition, some codes act as 'stop' and 'start' codes.

Example of DNA transcription

DNA code	AAT-CGT-AGG
mRNA codon	UUA-GCA-UCC
tRNA anti-codons	AAU-CGU-AGG

The example in the table above will code for three amino acids which will make up a small part of a peptide chain. A number of peptides linked together make a polypeptide and a number of polypeptides linked together make a protein.

The assembly of amino acids from these triplet codes, to form proteins, is described as the **translation** of the RNA code.

Protein synthesis

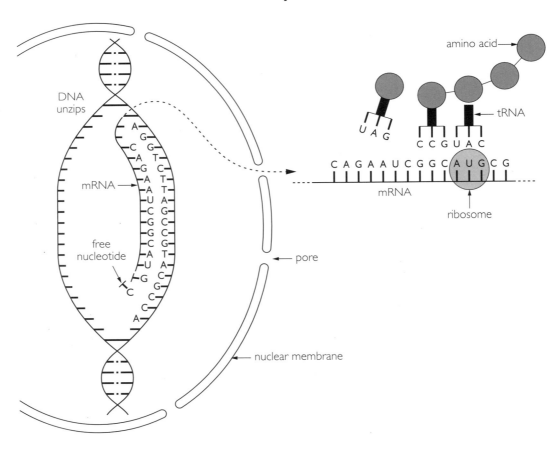

Energy Transfer

All living things respire, as it is this chemical process which liberates the energy in food to enable life functions such as growth, repair and movement to take place. The process is sometimes referred to as 'chemical' or 'tissue' respiration to distinguish it from its common usage to mean 'breathing'.

Respiration involves the **oxidation** of glucose to carbon dioxide and water, although other substances, such as fats, can also be oxidised. This transfers the chemical energy of the foodstuff to another energy-rich compound called adenosine triphosphate (**ATP**) which can be used in metabolic reactions as an immediate source of energy. If glucose solution is added to fresh muscle tissue nothing happens; if ATP is added, the actin and myosin fibres slide closer together and the muscle contracts.

ATP is constantly manufactured in all living cells from adenosine diphosphate (**ADP**) and inorganic phosphate (**P_i**). The rate of production of ATP varies to meet the demands set by the cell. So the mass of ATP in the body remains fairly constant at all times.

The process of respiration starts in the cytoplasm and continues in the mitochondria if oxygen is present. There are many small steps in this **aerobic** (with oxygen) process to ensure the energy is released in a gradual and controlled way. In essence, hydrogen is removed from glucose ($C_6H_{12}O_6$). This leaves an excess of carbon and oxygen atoms which are released to the atmosphere eventually as carbon dioxide. The hydrogen is then combined with oxygen from the atmosphere to make water. As water is formed, a large quantity of energy becomes available for the synthesis of ATP, so much so that 38 molecules of ATP can be produced from the complete oxidation of only one molecule of glucose.

If insufficient oxygen is available, the mitochondria are not involved; instead an **anaerobic** (without oxygen) process, which is much less efficient (net production of only two molecules of ATP), occurs. Moreover, the poison **lactic acid** is produced and this has to be removed later by the body, using oxygen. This happens after heavy exercise and is referred to as 'repaying the oxygen debt'.

Brief summary of chemical respiration

O – oxygen
H – hydrogen
CO_2 – carbon dioxide

* very active mitochondria have many cristae

Anaerobic and Aerobic Respiration

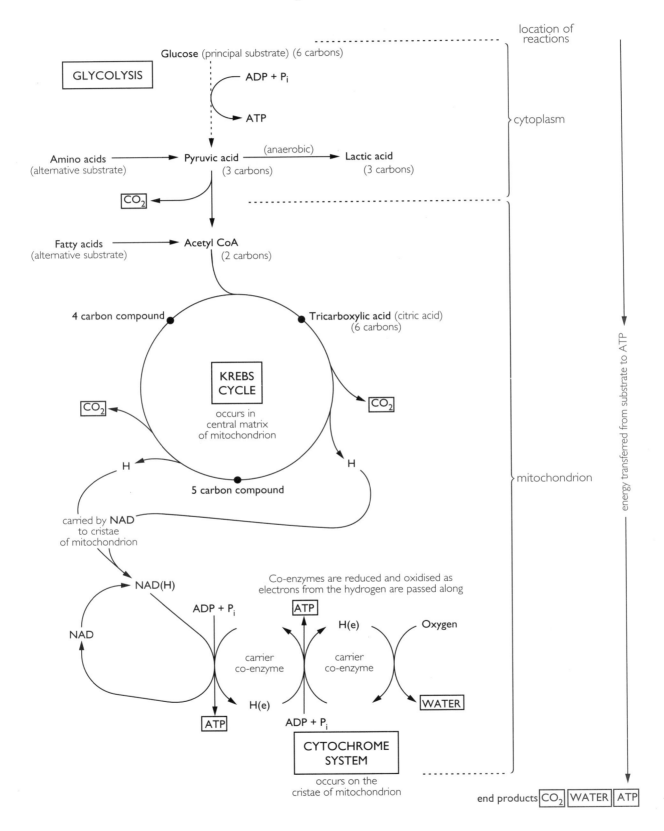

Sources of Energy

Metabolism is the sum total of all the chemical reactions taking place in the body and can be subdivided into two categories: (1) Anabolic reactions cause simple molecules to combine to make complex molecules and require the input of energy. (2) Catabolic reactions cause complex molecules to break down to simple molecules with the release of energy. All chemical reactions involve the transfer of energy.

A glucose molecule [$C_6H_{12}O_6$]

CH₂OH

Many compounds are used as sources of energy in the cell. Proteins, fats and carbohydrates can all be oxidised to produce ATP. However, glucose, a **monosaccharide carbohydrate**, is the most common respiratory substrate. Monosaccharides are the building blocks of all carbohydrates and are often called 'simple sugars'. They contain a number of carbon atoms combined with hydrogen and oxygen atoms in the ratio of 2:1, as in water (H_2O) – hence the name, carbohydrate. Their general formula can be summarised $[CH_2O]_n$ where n is the number of carbon atoms in the sugar.

The most important monosaccharides in the body are glucose, fructose and galactose (with 6 carbons each); deoxyribose and ribose (with 5 carbons each). Fructose and galactose are converted by the liver to glucose. Deoxyribose and ribose are components of DNA and RNA.

A disaccharide

If two monosaccharides are combined, a **disaccharide** is formed. The commonest disaccharide is table sugar (properly called **sucrose**) which is composed of glucose and fructose combined together. Milk sugar (**lactose**) is a disaccharide composed of glucose and galactose. Malt sugar (**maltose**) is a disaccharide composed of two molecules of glucose.

When many monosaccharides are combined together, large insoluble **polysaccharides** are formed, such as starch, glycogen and cellulose. These three are all composed of many glucose molecules combined in a variety of ways. Since these molecules are insoluble, they are very useful as stores of glucose because they have no osmotic effect on their surroundings. Highly concentrated solutions of glucose would draw water out of cells and destroy them. Humans, like many other animals, store glucose as **glycogen** in the muscles and the liver. Excess glucose can also be stored as fat (a **lipid**).

A polysaccharide

Lipids liberate twice as much energy as carbohydrates (per unit mass) but their breakdown takes longer. So in a marathon race – when glycogen stores are exhausted and only fats left as a source of energy – the runner 'hits the wall' and slows down to a crawl, totally exhausted, but not in fact out of energy. There is still plenty of fat available – it is just a slow process transporting and metabolising fat to generate ATP. Interestingly, the heart, unlike other muscles of the body, has a preference for fat as a source of energy. If the heart relied solely on glucose then the exhausted runner would have died! Experienced marathon runners load up with carbohydrate before the event by altering their diet, then pace the race so that there is just sufficient glucose and glycogen to maintain top speed to the finishing line.

A simple fat molecule (triglyceride)

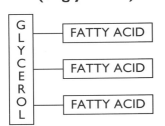

Lipids are similar to carbohydrates, in that they are composed of carbon, hydrogen and oxygen, but the atoms are present in different proportions. They include waxes, oils, fats and steroids. Like polysaccharides, they are very useful as stores of energy because they are insoluble in water. But they have many other important functions:

- Nerve fibres are insulated by a fatty myelin sheath.
- Our feet and hands are protected by pads of fat.
- Phospholipids are essential components of all membranes.
- Vitamins A and D, which are insoluble in water, are transported by fats.
- Sebum, a wax, waterproofs our skin and protects it from bacteria.
- Steroids, such as oestrogen and testosterone, are important hormones.
- Cholesterol, another steroid, is also essential for membrane construction.

Proteins are not stored in the body. So, when excess proteins are broken down, by deamination in the liver, some of the products can enter the Krebs cycle and be broken down to release energy. This provides a small proportion of our daily energy requirement, but is not important. However, in times of extreme starvation when carbohydrate and fat reserves have been almost used up, the body will start to break down its own structural proteins, such as actin and myosin, to obtain energy. Eventually, when the heart muscle is broken down, death results.

Cell Membranes

Cells are full of membranes and are surrounded by a membrane called the **plasma membrane**. Many organelles (such as mitochondria, endoplasmic reticulum and the nucleus) are bounded by, or composed of, membranes.

Membranes are exceedingly thin and are made of **phospholipids** and **proteins**. The lipids form two layers of molecules which are mobile. The proteins are found scattered as a mosaic in and on the lipid layers and they too can move around the membrane.

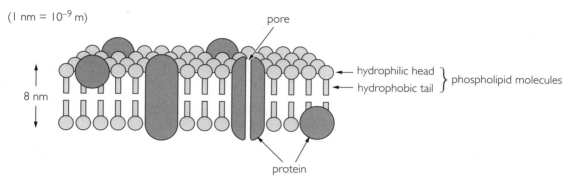

For this reason, the membrane is described as a **fluid-mosaic**. Membranes keep their shape because of the hydrophobic and hydrophilic nature of their lipid molecules.

One end of the molecule is repelled by water (hydrophobic) and the other is attracted to water (hydrophilic). As a consequence of their fluidity, membranes can often recover from minor physical damage. Lipids also allow small molecules, such as water molecules, to pass through the membrane unaided. The surface area of cells or organelles can be increased by the folding of membranes. Examples are the microvilli of cells lining the gut and the cristae of the mitochondria.

Because they are composed of proteins and lipids, membranes are easily damaged by heat, acids and by fat solvents such as alcohol. This can be demonstrated in the laboratory using cells from beetroot or red cabbage which have a coloured sap. When the membrane of these cells is damaged the coloured sap leaks out into the surrounding water.

Membrane proteins have many functions:

- There are enzymes present in membranes for many chemical reactions which take place on the surface of the membranes, e.g. stages in protein synthesis and respiration.

- There are receptor sites for hormones which then influence the activity of the cell.

- There are proteins which act as a skeleton for the membrane to give it shape and to allow it to move. For example, the membrane moves during endocytosis and exocytosis and when cilia waft to and fro.

- There are proteins which actively transport materials across the membrane using ATP as a source of energy – this is called **active transport**.

- There are proteins which form pores in the membrane through which substances can pass.

- There are proteins which act as markers for self-recognition.

Membrane Transport

Because cells are full of hundreds of different enzymes and millions of molecules which react every second, it is important that the internal environment of the cell is kept as constant as possible. In this respect, the plasma membrane plays a very important function by regulating what enters and leaves the cell. It allows some substances to pass through unaided, while others are transported actively and yet others are entirely prevented from passing through. For this reason membranes are said to be **selectively permeable**.

Diffusion

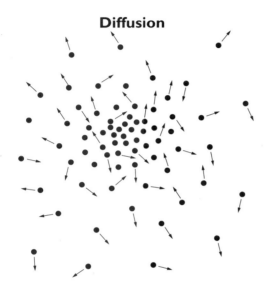

Atoms and molecules of gases and liquids move about at random. As a result they tend to spread themselves from areas of high concentration to areas of low concentration. This process is called **diffusion**. **Osmosis** is the term used to describe the diffusion of water across a membrane from a weak solution (high proportion of water molecules) to a strong solution (low proportion of water molecules). Cells placed in fresh water tend to burst because water diffuses into them. Cells placed in a strong salt or sugar solution shrink as water diffuses out of them. So it is important that our bodies maintain a constant osmotic environment in the tissue fluids.

Some membrane proteins allow molecules to pass through under their own 'power' by diffusion. But, if a cell needs to move a substance against the **concentration gradient**, it must supply energy in the form of ATP and transport the substance **actively**. There are proteins in membranes which act as pumps to force substances, such as mineral ions, from one side of the membrane to the other against the concentration gradient. Because **active transport** requires energy supplied by the cell, it is affected by the rate of respiration taking place in the cell. So factors such as the temperature, availability of oxygen and glucose, and the presence of enzyme inhibitors all affect rates of active transport.

Pinocytosis

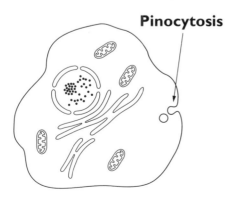

Large insoluble molecules (e.g. proteins) and larger items (e.g. whole bacteria) cannot pass through the membrane as such, but they can be transported from one side to the other in packages of membrane. The membrane flows round the item to form a tiny vacuole (vesicle) and the item is moved into (**endocytosis**) or out of (**exocytosis**) the cell. If large solid molecules are engulfed by the cell on a grand scale, the process is called **phagocytosis**. If fluid is engulfed by the membrane, the process is called **pinocytosis**.

Immunity

Throughout our lives, our bodies are constantly bombarded by foreign micro-organisms and the poisons they produce. To protect ourselves, a wide variety of defence systems have evolved. In the first instance, we keep organisms out of our bodies by a variety of **innate** (inborn) first line defence mechanisms. For example, the epithelial cells of the skin, the mucus membranes of the lungs and the gut, the ciliated cells of the respiratory tract, stomach acid, vaginal acid, skin sebum and the enzyme lysozyme of tears, all prevent, or reduce, the chances of invasion by foreign organisms. If germs break through these first line defences, the body fluids (blood, tissue fluid and lymph) contain many cells of the **immune system** which can neutralise invading microbes and their toxins in a variety of ways.

A foreign substance (usually a protein or polysaccharide) which triggers a reaction from the immune system is given the general name **antigen**. Any protein which the body produces in response to an antigen invasion is called an **antibody**. But not all foreign substances provoke an attack. If foreign molecules are fairly simple, the immune system does not recognise them as being foreign and leaves them alone. For example, plastic molecules, with many simple identical repeating units, can be used to manufacture artificial implants, such as hip joints and heart valves, because they do not stimulate an immune response.

Phagocytosis

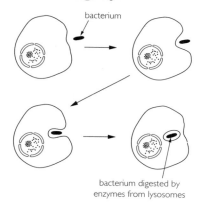

bacterium

bacterium digested by enzymes from lysosomes

The cells of the immune system are not organised as distinct organs like other body systems but, rather, consist of billions of cells throughout the body which are able to deal with antigens in one way or another. There are so many that if they were grouped together to form a single organ, they would occupy a volume similar to that of the brain. Some cells, **macrophages**, deal with invasion in a generalised non-specific way. Macrophages are phagocytic cells found throughout the body, but particularly in the lymph system. They engulf any foreign particles they find. Others are highly specific and each one can deal only with one type of antigen. These are the **lymphocytes**.

There are two basic types of lymphocyte, the **T-lymphocyte** and **B-lymphocyte**. Both types are white blood cells and look identical under the microscope but they mature in different regions of the body and act in quite different ways. The T-lymphocytes are so called because they mature in the thymus gland. There are a number of different types of T-lymphocytes with a variety of functions but, in essence, they attack and destroy body cells which have been infected by viruses or bacteria. These infected body cells can be identified because of markers (a carbohydrate + protein) on the plasma membrane which have been altered by the presence of the microbe. Every person has their own unique 'signature' (set of markers) on the surface of all their cells, except for identical twins who have identical markers.

During development of the fetus, the immune system learns to recognise these 'self' markers and thereafter recognises foreign antigens as being 'non-self'. A special binding site on the T-lymphocyte binds with the 'non-self' cell and destroys it, so destroying the microbe within. This is called the **cell-mediated response**.

The T-lymphocytes will also remove cancer cells and grafted tissue which they recognise as being foreign. For this reason, grafting of tissues or organs from one person to another is fraught with difficulty. Surgeons take care to match the antigens of each person as well as they can before transplant operations. Drugs are also used to suppress the immune system of the patient but, even then, donated organs are often rejected.

The B-lymphocytes mature in the bone marrow and protect the body by producing specific antibodies for specific antigens. The antibodies act at a distance from the parent B-lymphocyte and this is called the **humoral response**. The antibodies bind to antigens, making them temporarily inactive, and often act as markers for recognition by T-lymphocytes and phagocytic cells such as macrophages.

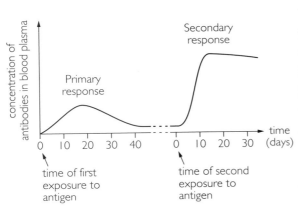

Very early in life we acquire millions of different T-lymphocyte and B-lymphocyte cells, ready-made, to deal with any specific antigen which might invade our bodies. Consequently, many T-lymphocytes and B-lymphocytes, which are never needed to fight infection, exist in the body. When an antigen invades, the appropriate T-lymphocytes and B-lymphocytes are already in place, but there is a delay until sufficient numbers of the same type are generated by mitotic division to deal with the invasion. However, once these cells have experienced a specific antigen, special memory cells linger in the body fluids and enable the immune system to produce more antibodies more quickly if the same antigen is experienced again. In other words, we become immune to the disease.

The immune system, like any other body system, can sometimes malfunction and fail to recognise body cells as 'self'. In such cases an '**autoimmune response**' occurs and the immune system attacks its own body cells as if they were foreign. There are a number of diseases which are thought to arise from autoimmunity, including forms of rheumatoid arthritis, diabetes and myxoedema.

Sometimes we suffer an **allergic reaction** to relatively innocent substances such as pollen grains, animal hair, feathers or food additives. Such substances are recognised as being foreign by the immune system which overreacts in an attempt to deal with the 'antigen'. About 30% of the UK population has an allergy of one kind or another, such as hay fever, eczema and food allergies. Children often grow out of allergic reactions, but an increasing number of adults suffer from allergies because they are bombarded by a growing cocktail of toxic chemicals in their daily lives.

The body can become 'immune' to antigens in a variety of ways, some of which occur naturally and some of which are artificially induced. The table summarises the various forms of immunity and how they are achieved.

Types of Immunity

Immunity	*Active*	*Passive*
Natural	antibodies made by body after an infection	antibodies passed across placenta or given in breast milk
Artificial	antibodies made by body after a vaccination	antibodies given by injection/ inoculation
advantages and disadvantages	slow acting but long-term protection	quick acting but only short-term protection

Viruses

Viral replication

virus →
bacterium →
DNA injected
DNA replicated
Protein coats manufactured
DNA incorporated into protein coat
Cell bursts and viruses released

Viruses are the smallest known living things (10–100 nm). In fact some would question whether they can be classified as living organisms. They cannot reproduce on their own, they do not grow, they do not respire, they do not feed and they do not produce waste. They can be crystallised like simple chemicals and they are invisible except under the most powerful electron microscopes. Despite their size, however, they can have profound effects on other living organisms because they are always parasitic. They make use of the biochemical machinery of host cells to make copies of themselves and, in so doing, destroy the host cells.

Viruses are made of a protein coat containing a DNA or RNA core. When they combine with host cells, they inject their DNA or RNA into the cell. The host cell then comes under the command of the new genetic instructions and makes new viruses using nucleotides, amino acids, ATP and enzymes supplied by the host cell. The viruses, once formed, burst out of the host cell, killing it in the process. For this reason we suffer symptoms of illness and may even die from a viral infection.

Many human diseases are caused by viruses, and such diseases are difficult to treat because they do not respond to the use of antibiotics. We have to leave it to our own immune system to deal with the viral attack. Human diseases such as polio, AIDS, smallpox, rabies, measles, mumps, the common cold and flu are all caused by viruses.

Inheritance

A karyotype of a male

sex chromosomes
X and Y
23

We are like our parents because we inherit their **genes**. Genes are pieces of DNA which code for every one of our characteristics. Each of our parents provides us with a complete set of genes via their respective sperm and egg. So all of our cells contain two sets of instructions for every characteristic of our body – one set from each of our parents. These pairs of alternative genes are called **alleles**. Our genes make up the **23 pairs** of **chromosomes** found in every body cell (**somatic** cell) except in sperm and eggs (**gametes**) and mature red blood cells. Gametes have only one set of chromosomes and red blood cells have none. A diagram to show these chromosome pairs is called a **karyotype**.

Before any cell divides, it must duplicate its DNA and ensure that each daughter cell contains a copy of the code. The process of copying the DNA is called **replication**, and the process of separating the copies of the DNA (chromosomes) and delivering them to each new cell is called **mitosis**.

Replication of DNA

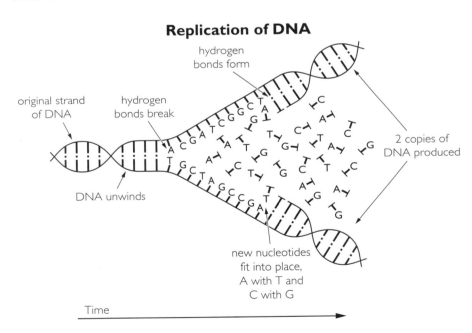

Mitosis

Chromosomes appear as pairs of **chromatids** and the nuclear membrane disappears. The chromosomes migrate to the equator of the cell where the chromatids separate and are pulled apart to the poles of the cell by spindle fibres. The nuclear membrane reforms and cell division follows. Each new cell has a complete set of copies of the original chromosomes.

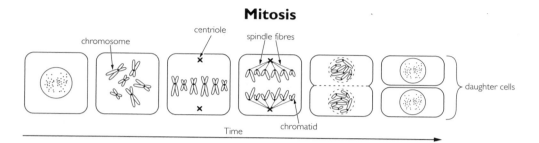

Mitosis

This means that all the cells of the body contain identical sets of genes. However, each somatic cell only uses those genes which are of relevance to it. For example, only red blood cells use the gene for the manufacture of haemoglobin. How this is achieved is not yet clearly understood.

Meiosis

All somatic cells contain **46** chromosomes. This is called the **diploid** number. However, gametes only contain **23** chromosomes, the **haploid** number, so that when a sperm fertilises an egg the full diploid number of chromosomes is regained. When gametes are formed, the process of nuclear division which reduces the diploid number to the haploid number is called **meiosis**. Also, during meiosis, genes are shuffled such that each haploid gamete which is formed has a new combination of genes. So, when the gametes fuse, the **zygote** (fertilised egg) which is formed has two unique sets of genes, one set from each parent. These matching sets of genes are on pairs of chromosomes called **homologous** chromosomes. The genes on each of these homologous pairs are found at the same points, called **gene loci**, on the chromosomes.

The shuffling of the genes during meiosis results from two different events:

1. the independent arrangement of pairs of chromosomes at the equator of the cell

2. the crossing over of pieces of chromosome at points called **chiasmata** (singular – 'chiasma').

This is shown in the diagrams over the page.
(For clarity only four chromosomes are shown.)

Meiosis

Note: chromosomes originating from sperm are shown as shaded; chromosomes from egg are shown as unshaded.

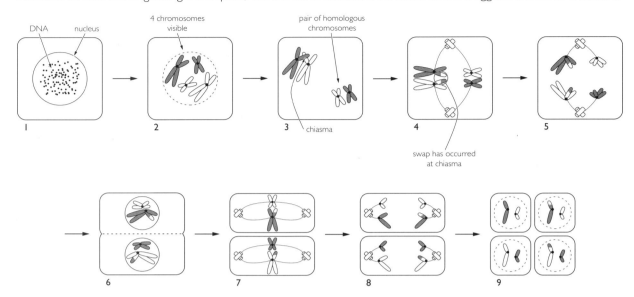

1 The chromosomes are normally invisible in the nucleus, as the DNA is not densely coiled. Before meiosis can take place, the DNA must replicate so that there are two copies of the genes to share out to the new cells which are about to be formed.

2 The chromosomes start to shorten and thicken and eventually become visible as tiny dark threads. They can be seen as pairs of **chromatids** joined at a point called the centromere. These chromatids are the identical copies of DNA formed during earlier replication.

3 The chromosomes find their homologous partners and the chromatids become intertwined at chiasmata. At these points the chromatids break and rejoin with their opposite number and genes **cross over** from one chromatid to another. This does not happen in mitosis.

4 The homologous pairs of chromosomes migrate to the equator of the cell and line up in pairs independently of one another. Spindle fibres form at the poles and become attached to the chromosomes at their centromeres.

5 The homologous chromosomes are pulled apart towards opposite ends of the cell by the spindle fibres.

6 Once the chromosomes have reached the poles of the cell, the cell divides in two. Now each new cell has only two chromosomes and not four.

7 The two cells go on to divide again to form four sperm cells or one egg cell. (The three other potential egg cells degenerate during the process.) This second meiotic division is very similar to a mitotic division. Chromosomes line up along the equator individually.

8 The spindle fibres form and become attached to the chromatids which they pull apart.

9 Four cells are formed, each with two chromatids which can now be called chromosomes again. Each chromatid will be different from every other chromatid because of the crossing over of genes at chiasmata. Moreover, the combination of chromatids in each cell will be different because of the random way in which they lined up along the equator of the cell at the first division.

Monohybrid Inheritance

Genuine examples of human characteristics controlled by a single pair of alleles (monohybrid inheritance) are uncommon. Some of the examples which follow are simplified for ease of understanding.

Every person has two alleles for every characteristic, one inherited from each parent. If just one of these alleles affects the growth and development of the individual on its own, it is said to be **dominant**. If an allele has no effect on an individual on its own, it is said to be **recessive**. Recessive alleles only have an effect on the development of an individual when a person inherits *two* recessive alleles, one from each parent. For example, **Huntington's chorea** is a particularly unpleasant condition which appears in middle age and results in early death. The allele which causes the condition is dominant. So a person with only one of these alleles will suffer from the condition and has a 50% chance of passing the allele on to his or her children. The condition **cystic fibrosis** is caused by a recessive allele. So people with *one* of these recessive alleles are perfectly healthy. However, if a man and woman *both* carry one of these alleles, there is a 25% chance that any child they have will be affected by cystic fibrosis. See the diagram below:

A^N = normal dominant allele
A^n = recessive cystic fibrosis allele

sperm / eggs	A^N	A^n
A^N	$A^N A^N$ NORMAL	$A^N A^n$ NORMAL but CARRIER
A^n	$A^N A^n$ NORMAL but CARRIER	$A^n A^n$ CYSTIC FIBROSIS

A number of technical terms are used to simplify genetic language. For example, if an individual has two identical alleles for any particular characteristic (trait) then they are said to be **homozygous**. Individuals with two different alleles are said to be **heterozygous**. The appearance of an individual with respect to a particular set of alleles is referred to as the **phenotype**. The genetic make-up of an individual is referred to as the **genotype**. Genotypes are written using letters of the alphabet. Capital letters indicate dominant alleles and small letters indicate recessive alleles. The parents are referred to as the **P** generation and thereafter the offspring are called the F_1 and F_2 generations, etc.

M = allele for melanin (skin pigment)
m = allele for no melanin (albino)

☐ = male   ◯ = female

P   [Mm] × (MM)

F_1 [Mm]   (Mm) × [Mm]

F_2 [mm]   [MM]

A family tree showing the inheritance of **albinism** in a fictitious family is shown opposite. Albinos lack the ability to make the pigment melanin which colours the skin, hair and iris. A man, who is a carrier of the recessive allele, marries a woman without the allele. They have two children, both of whom inherit the allele from their father. The chance of both children inheriting the allele is 25% ($\frac{1}{2} \times \frac{1}{2} = \frac{1}{4}$). The son does not marry but the daughter marries a carrier of the allele and they have one albino child and one unaffected child. The chance of two carriers having an albino child is 25%. The chance of two carriers having a child who does not carry the recessive allele is also 25%.

Sometimes recessive alleles are not totally hidden by dominant alleles and the alleles are then referred to as **incompletely dominant**. In other cases, both alleles have an equal effect on the phenotype of an individual and are described as **co-dominant**. However, the distinction between incompletely dominant alleles and co-dominant alleles is not easy to quantify. Some examples are given below:

Sickle-cell anaemia is a recessive condition in which blood haemoglobin is malformed. Anyone with two 'sickle-cell' alleles suffers from severe sickle-cell anaemia. But anyone with only one 'sickle-cell' allele and one 'normal' allele suffers from a mild form of the disease. This is because only some of the haemoglobin is affected. Interestingly, those with mild sickle-cell anaemia are more resistant to malaria. The disorder is much more common in areas of the world where malaria is endemic (always present) because someone with one 'sickle-cell' allele is more likely to survive a malarial attack than someone without the allele.

Red blood cells have a number of antigenic 'self' protein markers on their surfaces which are important in situations where blood has to be transfused from one person to another. If the protein markers of a donor's blood don't match those of the recipient, the recipient will recognise the donated blood as foreign and attack it with antibodies. Coagulation can then occur and the recipient can die. One such blood group system is the ABO system. The genetics of this system are slightly more complicated because three alleles are involved. They are described as **multiple alleles**. An individual can, of course, have only two of these alleles, one from each parent. Alleles A and B are co-dominant to allele O, which is recessive. The table summarises possible genotypes and phenotypes.

Blood groups (phenotypes)	Alleles (genotypes)
A	AO or AA
B	BO or BB
AB	AB only
O	OO only

People with blood group A have anti-B antibodies in their **plasma** (blood fluid), so they cannot receive blood from those of blood group B or AB. Likewise, people of blood group B cannot receive blood from those with blood group A or AB. People with blood group O are universal donors because they have no antigenic markers on their red blood cells; those of blood group AB are universal recipients (receivers of blood) because they have no antibodies against red blood cells in their plasma.

Another blood group system is the **Rhesus system**. Individuals who are Rhesus-positive have an antigenic marker on their cells; those who are Rhesus-negative do not. The allele for the antigenic marker is dominant.

Sex Linkage

In humans, the sex of a person is controlled by genes found on the sex chromosomes. There are two sex chromosomes, a longer one called the **X-chromosome** and a shorter one called the **Y-chromosome**. Women have two X-chromosomes; men have an X- and a Y-chromosome. The other 44 chromosomes are referred to as **autosomes** (see karyotype on page 20).

egg / sperm	X	X
X	XX	XX
Y	XY	XY

XX = 50% of offspring are girls
XY = 50% of offspring are boys

During meiosis the sex chromosomes separate so that eggs can only have an X-chromosome and sperm can have a Y-chromosome or an X-chromosome. It is the sperm therefore which dictates what sex a child will be.

Not all the genes on the sex chromosomes are involved in the development of sexuality. Many have nothing to do with sex at all. These genes are said to be **sex-linked** and their inheritance does not appear to follow normal monohybrid patterns because of the lack of such genes on the short Y-chromosome. Males cannot therefore be heterozygous for such sex-linked genes. Instead, they always express these genes in their phenotype regardless of whether the genes are dominant or recessive.

egg / sperm	X^c	X^c
X^C	$X^C X^c$	$X^C X^c$
no allele Y	$X^c Y$	$X^c Y$

c = allele for colour blindness

C = allele for normal vision

Two examples are the genes for colour vision, and a factor involved in the clotting of blood. Their mutant recessive alleles give **colour blindness** and **haemophilia** respectively. So a colour-blind woman, having children to a normal man, will have boys who are colour-blind and girls who, although carriers of the gene, have normal colour vision, as shown in the table opposite.

Polygenic Inheritance

Most characteristics of organisms, including humans, are controlled, not by one pair of alleles but by many pairs of alleles. This is referred to as **polygenic inheritance**. For example, if height in humans were controlled by a single pair of alleles, it would result in humans being either short or tall, rather like Mendel's pea plants. The feature would show **discontinuous** variation.

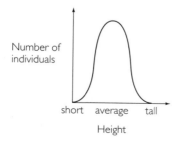

In fact, height is controlled by the combined effect of many alleles at different loci on the chromosomes. Height is also influenced by the environment, particularly by nutrition. So there exists a wide range of possible heights, and human height shows **continuous** variation. When heights or weights of a large population of humans of the same age are plotted on a graph they result in a line which shows a **normal distribution**.

Mutations

Meiosis, followed by random fertilisation, results in an almost infinite variety of gene combinations in any species of plant or animal which reproduces sexually. However, meiosis can only recombine genes which already exist – it cannot create new genes. New genes arise as a result of **mutations**.

Mutations can, and do, occur in any nucleated cells. However, only those mutations occurring in our gametes are of significance, because they are inherited. Mutations are random, undirected, spontaneous changes in DNA molecules, and can be as simple as a change in one base, or as significant as a change in large parts, or indeed numbers, of entire chromosomes in a cell.

Although mutations are the essential raw material for evolution, a vast majority are harmful. Only rarely does a mutation arise which gives an organism an improved chance of survival in a constantly changing environment. For example, the mutation in humans giving rise to haemophilia is harmful, but the mutation giving rise to sickle-cell anaemia gives heterozygotes some protection from the disease malaria.

Mutations

AUG/CCC/GGC/AAU	normal mRNA
AUG/*CUC*/GGC/AAU	substitution (U for C)
AUG/*CCG/GCA/AU...*	deletion (take out C)
AUG/*CCA/CGG/CAA/U*	insertion (add A)
(affected codes in italics)	

There are many different ways in which mutations can occur. Individual bases can be added, substituted or deleted from a strand of DNA. **Substitutions** often have little effect because they change only one triplet code and might affect only one amino acid in a protein chain. However, **deletions** and **insertions** are much more serious because they alter the subsequent code for every amino acid in the chain. (See table.)

Chromosome mutations can result in pieces of chromosome being removed (**deletion**), rotated and replaced (**inversion**), shifted from one place to another (**translocation**), or copied twice (**duplication**).

Occasionally, entire extra chromosomes arise in gametes through a failure of the spindle fibres to pull one or more of the chromosomes apart at meiosis. This meiotic failure is called **non-disjunction**. Most of these mutations are lethal but a few are not. For example, an extra autosome number 21 causes **Down's syndrome**. An extra X-chromosome in males (XXY) causes sterility, an increase in height and the development of some female characteristics such as breasts. This is called **Kleinfelter's syndrome**.

Anything which causes a mutation is called a **mutagen**. The shorter wavelengths of radiation are particularly dangerous. Alpha (α) rays, beta (β) rays, gamma (γ) rays and X-rays, which are emitted from nuclear reactions and atomic explosions, increase mutation rates, and can be lethal if of sufficient intensity. The mutation rate in humans has increased in the last sixty years due to the testing of atomic bombs and the operation of nuclear power stations. Pregnant women are rarely X-rayed, to avoid causing mutations in the developing embryonic tissues of the fetus. Moreover, people working with X-ray machines have to be protected from the dangerous radiation. UV light and many toxic chemicals are also mutagenic. Cigarette smoke causes cancer of the lungs and sunbathing increases the risk of skin cancer. Cancer results from mutations arising in somatic cells, where these mutated cells divide and multiply in an uncontrolled way. If malignant, these cells spread throughout the body and cause new cancerous growth in other tissues.

Genetic Screening and Counselling

The word 'disease' is used to describe many human conditions, but it is perhaps best reserved for those conditions which have been caused by **pathogens** (organisms which cause disease), such as viruses and bacteria, rather than those caused by faulty genes. Diseases caused by genes are perhaps better described as **genetic disorders**. Genetic disorders have a permanence about them which makes them difficult to treat; and, if there is a risk we have inherited a defective gene or genes, then we may wish to know what that risk is.

There are a number of ways of determining risk and much depends on the number of genes which affect the condition and on the condition itself. At its simplest, if only one gene is involved, as in haemophilia or cystic fibrosis, then the risks are easy to evaluate. But, if many genes are involved, the calculations become much less precise. People who advise couples on such risks are called **genetic counsellors**. They rely on family histories, and the study of cells, chromosomes and DNA, to make their predictions. For example, if a man has a father who is a haemophiliac (haemophilia is a sex-linked recessive condition), then a counsellor can tell him that he could not have inherited the gene from his father. On the other hand, a woman whose father has haemophilia will be a carrier and may pass the allele to her sons or daughters.

One form of Duchenne **muscular dystrophy** is inherited in the same way as haemophilia except that, as sufferers die young, the condition is only passed down the female line. Screening tests to detect this gene before birth can be unreliable, as a third of all cases result from new mutations and are therefore born into families with no previous history of the disorder. The physical symptoms do not appear until after the age of 2 or 3. However, a family history of muscular dystrophy enables counsellors to make predictions of risk. A carrier woman has a 50% chance of having a carrier daughter or an affected son, and the daughter has a 25% chance of having an affected son.

Predictions involving polygenic disorders are much more difficult; in these cases the risks are evaluated by looking at family histories. At its simplest, if your father and grandfather died of a heart attack before the age of sixty, then your own life is at greater risk. Genetic predisposition to various disorders is an important factor when considering life expectancy, and life insurance companies take these factors into account when assessing whether an applicant is a safe risk. It is thought that conditions such as cancer, diabetes and heart disease are all influenced by genetic, as well as by environmental, factors.

If parents suspect their child has a genetic defect then, in some cases, the presence of the defective allele can be detected using a variety of genetic screening tests. For example, amniotic fluid containing embryo cells can be removed by **amniocentesis**. The cells of the embryo are then grown and examined for DNA and chromosome defects. Karyotypes can be examined for chromosome abnormalities, such as the one which causes Down's syndrome. Such tests carry an element of risk themselves, and so are not routinely carried out.

Amniocentesis

Another method of screening is to examine the blood cells or urine of a person. The disorder phenylketonuria (PKU) is checked for in this way. (See page 7.)

As genetic screening becomes more effective, and techniques in gene therapy improve, it seems likely we will eventually be able to cure many genetic disorders. But in doing so, we all face many difficult questions with regard to the ethics of tinkering with our genetic inheritance.

UNIT TWO
THE CONTINUATION OF LIFE

This forty-hour Unit contains information on reproduction, birth, growth and development, influence of hormones, circulation, delivery and removal of materials to and from cells, and on regulating mechanisms.

Reproduction — Female

Girls become sexually mature at **puberty**, at around the age of 10 to 14, and at that time their monthly **oestrous (menstrual) cycle** begins. This cycle will continue until the girl reaches the age of around 50, when it stops permanently. This 'change of life' is called the menopause. Normally, the only other occasion when the oestrous cycle stops is when a woman becomes pregnant. Then the cycle is held in check for the nine months of **gestation**, until the birth of the child. The cycle lasts for around four weeks and is controlled by changes in concentration of four hormones: **follicle-stimulating hormone (FSH)**, **luteinising hormone (LH)**, **oestrogen** and **progesterone**. The first two are manufactured in the **pituitary gland**, which is attached to the underside of the brain, and the other two are produced by structures in the ovaries.

The cycle is initiated by the production of FSH which has two effects: (1) it causes an egg to mature and a ball of cells, called the **Graafian follicle**, to develop round the egg (ovum) in one ovary. A woman has thousands of eggs present in her ovaries from before birth, and during her life around 400 of these will mature to be released once a month. (2) FSH also stimulates the follicle to produce oestrogen. Oestrogen in turn inhibits the production of FSH, so keeping its production in check. This is an example of **negative feedback control**. Negative feedback mechanisms are commonly employed by the body to control processes, or concentrations of various substances.

Oestrous cycle

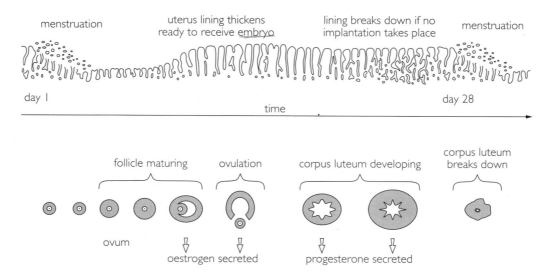

menstruation   uterus lining thickens ready to receive embryo   lining breaks down if no implantation takes place   menstruation

day 1   time   day 28

follicle maturing   ovulation   corpus luteum developing   corpus luteum breaks down

ovum   oestrogen secreted   progesterone secreted

Oestrogen also has a number of other effects on the body; it initiates the repair of the uterine wall, in preparation for a developing embryo, and stimulates the pituitary gland to produce a sudden upsurge of LH. The surge of LH stimulates **ovulation** – the rupture of the follicle in the ovary and the release of the egg. LH also stimulates the development of the follicle into the **corpus luteum**, a small 'yellow body' in the ovary.

The corpus luteum produces progesterone for a few days before degenerating if the egg has not been fertilised. The progesterone which it produces during this short time encourages further development of the uterine lining in preparation for a new embryo. High concentrations of progesterone are maintained in the body if fertilisation occurs, due to the survival of the corpus luteum for a number of months. When it does finally disappear, the production of progesterone is taken over by the placenta until birth. The progesterone helps maintain the lining of the uterus and also, importantly, inhibits the pituitary gland and thereby holds ovulation in check. It would not make biological sense to have another fertilised egg arriving at the uterus when one was already there developing. This is one of the reasons progesterone is the principal ingredient in most contraceptive pills. Oestrogen too is often used to help keep the menstrual cycle in check. If fertilisation does not occur in a normal cycle, the levels of progesterone fall, allowing menstruation to take place.

Reproduction — Male

Cross-section through seminiferous tubules

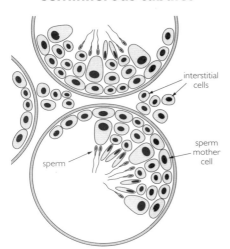

interstitial cells

sperm mother cell

sperm

The manufacture of sperm occurs in the seminiferous tubules of the testes. The development of male characteristics is influenced and controlled by a variety of hormones, similar to those present in females.

Testosterone is produced by the **interstitial cells** of the testes and influences the onset of puberty, the development of male sexual characteristics, and the development of sperm in the testes. Unlike most hormones, it is not a protein or peptide but a steroid (a lipid-like substance derived from cholesterol). Testosterone boosts basal metabolic rate and for this reason men tend to have a higher metabolic rate than women. The secretion of testosterone is influenced by two other hormones, **follicle-stimulating hormone** (**FSH**) and **luteinising hormone** (**LH**) which are produced by the pituitary gland. FSH, along with testosterone, indirectly stimulates the production of sperm in the testes. LH stimulates the interstitial cells to produce testosterone. In males, LH is therefore sometimes called, more logically, interstitial cell-stimulating hormone (ICSH). Testosterone has a negative feedback effect on the pituitary gland and thereby concentrations of FSH, LH and testosterone are controlled.

Infertility

Couples having regular intercourse at the fertile period (a few days either side of ovulation) can only expect a 30% chance of fertilisation in any month, and around 10% of couples are unable to conceive within a period of two years. The most common causes of infertility and their treatment are outlined in the table below:

Type of infertility	Possible causes	Possible treatment
failure to ovulate	hormone imbalance, ill health*, emotional stress	drugs which stimulate production of FSH and LH
blockage of uterine tubes	infection, tissue growth (e.g. tumour)	surgery, **in vitro fertilisation**†
failure of implantation	hormone imbalance	drugs which stimulate production of FSH and LH
low sperm counts	hormone imbalance, ill health*, emotional stress	hormone treatment, in vitro fertilisation, **artificial insemination**‡

* obesity, excessive alcohol consumption, excessive smoking, drugs and a variety of venereal diseases can all affect male and female fertility. Prolonged use of some contraceptives can also lead to permanent fertility problems.

† in vitro fertilisation is fertilisation in a glass dish. Sperm and eggs taken from the two potential parents are allowed to fuse 'in glass' in the laboratory. The zygote is allowed to divide a few times then replaced in the uterus of the woman to develop normally.

‡ artificial insemination is the insertion of sperm (obtained by masturbation) into the vagina, using a mechanical device such as a syringe. The technique is used when a man's sperm count is low or when his sperm are not viable (healthy) and another person's sperm are used.

Contraception

Contraception is the prevention of fertilisation or implantation by some means or other. There are many different methods: some mechanical, such as the cap and condom; others chemical, such as spermicidal creams. Two 'biological' methods are described below:

Form of contraception	How it works	Success rate
contraceptive pill – taken daily	contains progesterone, and sometimes oestrogen, which prevent ovulation	high
rhythm method – intercourse restricted to infertile period	infertile period is determined by checking temperature and vaginal mucus. Small rise in temperature (0·5°C) and thinning of vaginal mucus indicates ovulation	low

Prenatal Development

When a sperm fertilises an egg (in the oviduct), a **zygote** is formed. If two sperm fertilise two eggs, then **dizygotic** (fraternal) twins are formed. If one sperm fertilises one egg and the zygote later splits into two separate sets of cells, **monozygotic** (identical) twins are formed. After fertilisation, the zygote starts to divide as it continues on its way down the oviduct towards the womb. The first few divisions are called **cleavage** divisions because the cells simply split into two without any increase in size. As the cleavage divisions take place, cell size is continually halved until 'normal' cell size is achieved. The ball of cells arrives in the uterus a few days after fertilisation and eventually embeds itself in the wall (**endometrium**) of the uterus. This process is called **implantation** and usually occurs around one week after ovulation.

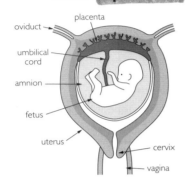

Shortly after implantation the embryonic cells start to specialise (**differentiate**) and form the organs of the body. By the end of the embryonic period, when the embryo is eight weeks old and 2 cm long, all the adult organ systems are recognisable. After this point the embryo is referred to as a **fetus**. The fetus is genetically different from its mother and would be treated as a foreign invader were it not for the fact that the mother's immune system is specially adapted to cope with the presence of the fetus' tissues.

Sometimes, however, the system breaks down. For example, if a Rhesus-negative (Rh−) woman has a Rhesus-positive (Rh+) child, there is a risk that she will produce antibodies against the Rhesus antigens on the red blood cells of her baby. This is not a problem with the first child because the response is too slow. But, if she has a second Rhesus-positive child, then there is a greater risk that antibodies will cross the placenta and destroy the child's blood cells. If a child is at risk, the standard treatment is 'exchange blood transfusion': Rh+ blood is removed from the baby and replaced with Rh− blood. Another more recent treatment is to inject the mother with anti-D factor which neutralises any Rh+ blood cells reaching the mother's blood. This prevents her from becoming sensitised and mounting an antibody attack.

The fetus receives food and oxygen from its mother and gets rid of waste, such as urea, through the **placenta**. In the placenta, the blood of the fetus and mother come into close contact but do not mix because the bloods may be of different blood groups, and because the mother's blood pressure would be too high for the fetus. Oxygen and carbon dioxide move across the placenta by **diffusion**, glucose by **active transport** and antibodies by **pinocytosis**. Unfortunately, many harmful substances can also pass from mother to fetus during this time. Alcohol, nicotine, many drugs, antibodies and viruses (e.g. rubella) can all pass across the placenta into the blood of the fetus and cause serious damage.

Oestrogen and progesterone are produced by the placenta towards the end of pregnancy and, as well as maintaining the endometrium, they stimulate the pituitary gland to produce a hormone, **prolactin**, which promotes the manufacture of milk by the breasts.

Birth

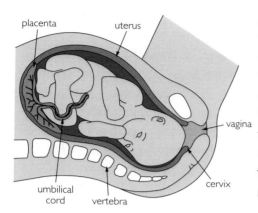

placenta
uterus
vagina
cervix
umbilical cord
vertebra

As the fetus increases in size towards the end of **gestation**, the uterus and cervix are stretched. This stretching process causes impulses to be sent to the **pituitary gland** where there is a release of the hormone **oxytocin**. As oxytocin concentration rises in the blood, it causes the contractions of 'labour'. Smooth muscle (involuntary muscle) in the walls of the uterus contracts in rhythmic waves, pushing the fetus out of the womb. The first sign that birth is about to happen is the flow of amniotic fluid from the vagina. Normally, a few minutes after the birth, the placenta is expelled.

Birth represents quite a shock to the infant. During the birth process, it is suddenly cast out of a warm watery environment into the cold, and its placental life support system is cut. Now the infant must do for itself all that the mother's body has been doing for it for the past nine months of gestation. It must breathe, feed, excrete and try to maintain its body temperature. Breathing is particularly difficult for premature babies who lack an important protein in their lungs which reduces the surface tension of water. Without this protein, the alveoli in the lungs collapse. Premature babies also lack fat stores.

Lactation is the production of milk by the breasts. For the first few days after birth, true milk is not produced. Instead, a yellowish fluid called **colostrum** is produced. Colostrum has a lower lactose content than milk and contains almost no fat, but it does have more protein, minerals and more vitamin A than true milk. Colostrum is particularly rich in antibodies which give the infant a degree of **passive immunity** against disease. However, breast milk can also contain poisons which the mother has unwittingly ingested in her food. Some persistent pesticides, such as DDT, which enter food chains can be detected in human breast milk.

After birth, the suckling of the infant induces the secretion of the hormone **prolactin** by the pituitary gland. Each time it suckles, impulses travel from the nipple to the **hypothalamus** which stimulates the pituitary to produce prolactin and oxytocin. The effect of oxytocin is almost immediate, in that it stimulates the ejection of milk from the glands in the breasts. Oxytocin also causes the uterus to contract to normal size. The prolactin surge stimulates the breasts to manufacture more milk for the next feed, and also inhibits the activity of the pituitary gland in the production of FSH and LH. So, when a baby suckles it not only obtains a rich supply of food but also prepares the mother to provide for the next meal and ensures that no other competitors are going to come along in nine months' time to share this unique supply of nourishment. Suckling, and the oxytocin response, are examples of a positive feedback mechanism, because the more the child suckles, the more milk it gets. The hormone oxytocin is sometimes given to pregnant women to induce birth, to reduce bleeding after birth or to stimulate milk production.

Postnatal Development

Average height increase per year

Growth is most rapid before birth as a fetus can double its weight in a matter of weeks during early development. However, after birth, growth is still relatively rapid and declines slowly until it ceases altogether at around 20 years of age. There are variations though, the most marked of which is the growth spurt which takes place at **puberty**. This is earlier in girls than in boys and is marked not only by an increase in size but also by many other physical changes:

Changes in females at puberty	Changes in males at puberty
increased growth of hair on body	increased growth of hair on body and face
widening of hips	increased growth of skeleton and muscles
maturation of sex organs	maturation of sex organs
eggs start to mature – one per month	sperm are manufactured – 1000s per second
start of oestrous cycle	deepening of voice

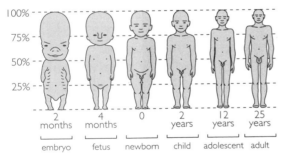

Growth progresses after birth and body proportions change, as shown in the diagram. At birth, the head is large, relative to the body, because the brain has to be well developed to cope with the large degree of learning which takes place during the early years.

Much of growth and development is controlled by a variety of hormones, the most important of which are given in the table below:

Hormone	Effect of hormone
growth hormone	promotes growth of bones and protein metabolism
thyroxine	controls cell growth and differentiation, and BMR*
oestrogen/progesterone and **testosterone**	promote the development of secondary sexual characteristics as described in the table above

* BMR = Basal Metabolic Rate. This is a measure of the basic chemical turnover of the body when at rest.

Transport Systems

All living things have a need to move substances around their bodies and the need for transport starts at cell level. All cells have to move substances round the cytoplasm and to and from their immediate environment. To do this they rely on a variety of methods, some of which have been studied already. In liquids and gases the ability of atoms and molecules to move on their own from areas of high concentration to areas of low concentration is called **diffusion** (see page 16). Our bodies make much use of this process, but the length of time it takes for a substance to diffuse from one place to another is proportional to the square of the distance. Put another way, if it takes one second for a molecule of glucose to diffuse across a cell, it will take over 50 years for the same molecule to diffuse a distance of one metre. Quite clearly, although diffusion is sufficiently quick to transport substances across cells, it is not sufficiently quick to get the glucose from the intestines to the active muscles of the arms and legs! So our bodies have developed other methods to increase the speed at which substances can be moved from one place to another. **Peristalsis**, breathing movements and the beating of the heart all aid the mass flow of materials round the body.

cell size

0·5 mm
0·5 mm
0·5 mm

body size

500 mm
500 mm
500 mm

Another consequence of great size is the change in the relative size of surface area to volume of any object. A tiny cell has a huge surface area in relation to its volume, whereas we humans have a relatively small surface area compared to our volume. This can be shown by a simple piece of arithmetic in the table below.

The small cube represents a very large cell, and the large cube represents our own body volume. From the table it can be seen that moving from cell size to body size, the surface area:volume ratio drops one thousand fold. So, quite clearly, not only is diffusion too slow to allow substances to move from one end of the body to the other but also the surface area of our body must be increased considerably to allow substances to enter and leave it by diffusion. For this reason we have, for example, lungs with a surface area close to that of a tennis court to ensure adequate gas exchange.

Dimensions	Cube	
	side of 0·5 mm	side of 500 mm
Surface area	0·25 mm²	250 000 mm²
Volume	0·125 mm³	125 000 000 mm³
SA ÷ Volume	2	0·002

The Circulatory System

The circulatory system consists of a four-chambered heart which pumps blood round the body. The blood travels through three types of blood vessel: arteries, veins and capillaries.

Arteries

thick muscular wall
(waterproof)

- have thick waterproof muscular walls
- pulsate due to the pumping action of the heart
- carry blood away from the heart to all the body organs
- have no valves
- contain oxygenated blood (except for the umbilical and pulmonary arteries)
- tend to lie deep in the body for protection
- divide into smaller arteries called **arterioles**

Veins

valve

thin wall
(waterproof)

- have thin waterproof walls with very little muscle
- carry blood at low pressure towards the heart from the body organs
- do not pulsate
- have valves to prevent the back-flow of blood
- usually contain deoxygenated blood (except for umbilical and pulmonary veins)
- are not so well protected as arteries
- are divided into smaller veins called **venules**

Capillaries

thin wall
one cell thick
(permeable)

- join arterioles to venules
- have walls one cell thick
- are permeable to allow exchange of materials between blood and tissues
- have no valves

The Circulatory System

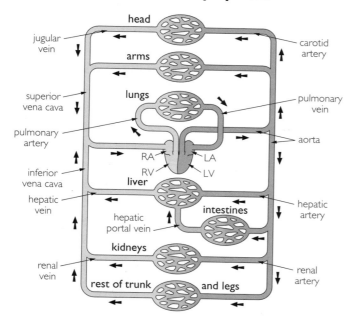

Blood consists of a fluid called the plasma, with red blood cells, white blood cells and platelets suspended in it. The composition of the blood is very precisely regulated by the lungs, liver and kidneys. The main arteries and veins of the body are shown in the diagram opposite.

The role of white blood cells – lymphocytes and macrophages – has been outlined earlier.

Platelets are tiny fragments of cytoplasm involved in the clotting action of blood.

The Heart

The average resting heart rate for adults is around 70 beats per minute, although fit athletes have resting heart rates below 40 bpm. This is because their heart chambers become larger and stronger as a result of regular training. During extreme exercise, a young adult's heart rate can rise to over 200 bpm. The heart rate of a fetus and young baby is much higher than 70 bpm and decreases as adult size is attained. It must be emphasised that these are average figures and there is considerable variation from one person to another. Stress, exercise, cigarette smoke and even the intake of food all have the effect of increasing heart rate.

The Heart

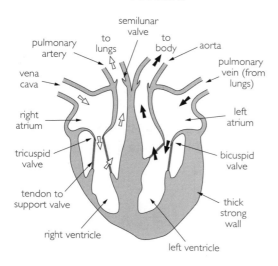

⇨ deoxygenated blood
➡ oxygenated blood

The heart has a unique form of muscle fibre which can contract every 0·8 seconds for the entire life of an individual. The fibres contain **actin** and **myosin** and are moderately striated. However, unlike striated muscle, the fibres are shorter, are usually uninucleate (one nucleus per fibre) and are interconnected. The links between the fibres allow them to contract at the same time. About 35% of the volume of cardiac muscle fibres is taken up by large mitochondria. The cardiac muscle fibres rely on aerobic respiration almost exclusively and use fatty acids most effectively, unlike other muscles of the body which use glucose as a primary fuel. The **coronary artery** supplies the heart muscle with blood.

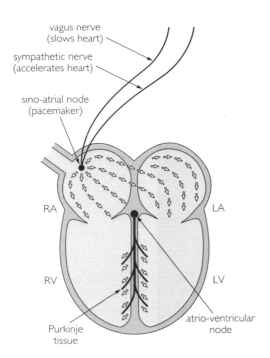

vagus nerve (slows heart)

sympathetic nerve (accelerates heart)

sino-atrial node (pacemaker)

RA

LA

RV

LV

Purkinje tissue

atrio-ventricular node

Most muscles contract as a result of impulses reaching them from nerves. This is not true of cardiac muscle which will continue to beat rhythmically even after its nerve supply has been cut. In the wall of the right atrium (auricle) is an area called the **pacemaker**, or **sino-atrial node** (**SAN**), which generates rhythmic electrical waves which travel across the walls of the atria, making them contract. Between the atria and the ventricles is another area called the **atrio-ventricular node** (**AVN**) which transmits the impulse from the atria to the ventricles so that they contract 0·1 seconds after the atria. A network of fibres (Purkinje fibres) spreads through the ventricles to carry the impulse from the AVN to all the cardiac muscle fibres. The contracting phase of the cardiac cycle is called **systole** and the relaxing phase is called **diastole**.

The pacemaker, if left on its own, would regulate the heart beat at around 70 to 80 beats per minute. However, the pacemaker is overridden by nerve stimulation from the **autonomic nervous system** via two nerves – the **sympathetic nerve** accelerates heart rate and the **parasympathetic nerve** (or **vagus nerve** as it is sometimes called) slows it down. In addition, the hormones **adrenaline** and **thyroxine** speed up heart rate. The autonomic control centre is found in the **medulla oblongata** of the brain and, to enable it to respond appropriately, it receives information from other sensors in the body which detect blood pressure, temperature, oxygen and carbon dioxide concentrations.

ECG trace taken during exercise

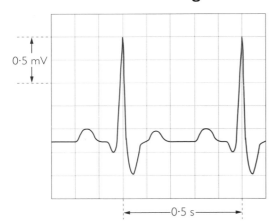

0·5 mV

0·5 s

The heart beat can be measured by the pressure wave which spreads through the arteries and is called 'pulse'. The heart can also be heard to beat using a stethoscope, the double 'lub-dub' sound arising from the closing of the two atrioventricular (**bicuspid** and **tricuspid**) valves followed by the two **semilunar** valves.

An important technique to determine heart function is to measure the electrical changes which take place using an **electrocardiograph** (**ECG**). In the example opposite, the heart rate can be calculated as 120 bpm.

Tissue Fluid and Lymph

Because blood flows in tubes of one kind or another, our circulatory system is said to be a closed system. At any point in the system, blood will flow from 'A' to 'B' so long as the pressure at 'A' is greater than the pressure at 'B'. It must follow, therefore, that blood pressure in the aorta is the highest, and blood pressure in vena cava the lowest, in the circulatory system. Pressure in the arteries fluctuates with the beating of the heart and is called **pulse**. Pulse and blood pressure are measured on arteries and not veins.

Blood pressure is measured using a **sphygmomanometer**. This is an instrument which measures two pressures: (1) the pressure required to stop blood flowing down the brachial artery in the arm – the systolic pressure – and (2) the pressure at which the blood is just able to flow freely along the brachial artery – the diastolic pressure. The pressure is measured in millimetres of mercury and is usually around 120/75 mmHg for a young adult. It tends to increase with age due to atherosclerosis (thickening of the artery lining), and can vary considerably from one person to another, and from minute to minute throughout the day.

Blood pressure in vessels supplying muscles at rest

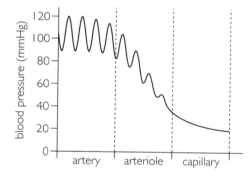

As the blood flows through the arteries there is little drop in pressure because the arteries have wide channels and elastic walls which offer little resistance to the flow of blood. However, when the blood reaches the much smaller arterioles, the pressure drops considerably. The blood pressure in the arterioles drops because there are many of them and they have a much smaller bore than the arteries. In effect this increases the surface area of the inner walls. It is the friction of the blood fluid in contact with the walls which reduces the pressure and the flow of the blood.

By the time the blood reaches the capillaries, the flow rate and pressure have dropped considerably. This is a desirable outcome because low pressure reduces fluid loss through the capillary walls and low flow rate allows time for exchange of materials to take place between the plasma and the tissues.

The blood pressure in the pulmonary circulation is five times lower than that in the systemic (general body) circulation. Since the blood does not have far to travel through the lungs this is not surprising. However, there is another important reason for the low pressure. The blood in the lungs must travel through delicate thin-walled capillaries in intimate contact with the air. If pressure were too high, fluid would leak out of the capillaries and fill the alveoli.

Not surprisingly, the higher pressure in the systemic circulation does cause fluid to leak out of the permeable capillaries into the surrounding tissues. Although some of this lost fluid returns to the blood at the venous end of the capillary system due to osmosis, a proportion does not return to the blood. This lost fluid is collected by the **lymph system** which is found in almost every tissue except for bone, teeth and the central nervous system (CNS). Every twenty-four hours, around three litres of this **tissue fluid** is collected by the tiny blind-ended lymph capillaries called **lymphatics**. Pressure in the lymphatics is even lower than pressure in the tissue fluid, so tissue fluid gradually flows into the lymphatics and moves along the lymph vessels to return to the main circulation at a point very near the heart.

Relationship between lymph vessels and capillaries

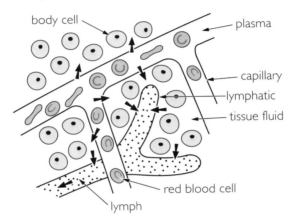

Lymph vessels, like veins, have valves which help maintain the flow of fluid so long as the body is moving. This tissue fluid in the lymph vessels is now called **lymph**. As it flows back towards the heart it flows through many **lymph glands** (**lymph nodes**) where harmful particles such as bacteria are removed or inactivated by the action of **macrophages** and **lymphocytes**. Remember that lymphocytes originate in the bone marrow but migrate to the lymph glands where they divide further and become active. Lymph glands swell up during times of illness (e.g. tonsilitis) as a consequence of their increased activity. The spleen and tonsils are part of the lymph system.

As a consequence of the flow of fluid in the circulatory system described above, plasma, tissue fluid and lymph differ in their composition. Most plasma proteins are too large to pass through the walls of the capillaries, so there is much less protein in the tissue fluid than in the plasma. For the same reason, red blood cells cannot escape from capillaries into the tissues but white blood cells are capable of squeezing between gaps in the capillary walls. Also, lymph tends to have more fat suspended in it because it is the specialised lymphatics of the villi, called **lacteals**, which collect digested fat and deliver it to the main circulation.

Red Blood Cells

Red blood cells are not true cells, in so far as they do not contain a nucleus and have few organelles in their cytoplasm. Mitochondria are missing because the principal role of red blood cells is to transport oxygen. It would be inefficient to have mitochondria using up the oxygen en route from the lungs to the respiring tissues of the body.

Red blood cell

⟵ 10 μm ⟶

Red blood cells are very small and **biconcave** in shape. Both these features have the effect of increasing the surface area of the cells relative to their volume. This means that oxygen can enter and leave the cells readily. What is more, oxygen does not have far to travel once inside the cell.

Mature red blood cells are full of **haemoglobin**, a red protein which has a high attraction (affinity) for oxygen. The small size of the corpuscles and the flexibility of their membranes also allow them to squeeze through the narrowest of capillaries.

Red blood cells are manufactured in the bone marrow and live for around 120 days. Worn-out red blood cells are destroyed in the spleen and liver. **Iron, vitamin B$_{12}$** and **amino acids** are the principal requirements for the manufacture of red blood cells and the iron is broken down by the body after the breakdown of the red blood cells. Haemoglobin is degraded to **bilirubin** in the liver and excreted in the **bile**. Anyone suffering from a shortage of red blood cells or haemoglobin is described as anaemic.

Gas Transport

The total pressure exerted by a mixture of gases is the sum of the pressures exerted independently by each gas in the mixture. The pressure exerted by each gas is called its **partial pressure** and is directly proportional to its percentage in the total gas mixture.

Since atmospheric pressure will support around 760 mmHg (millimetres of mercury) and, since oxygen occupies around 20% of fresh air, then the partial pressure of oxygen in fresh air is just over 150 mmHg. In the lungs there is always a lower percentage of oxygen than there is in fresh air. This is because inflowing air is mixed with stale residual air which is low in oxygen. So the partial pressure of alveolar air can be as low as 100 mmHg. However, this low figure presents no problem to haemoglobin which can become almost 100% saturated with oxygen at a partial pressure as low as 90 mmHg. When the haemoglobin reaches the respiring tissues, the partial pressure of oxygen is much lower there because the tissues are using it up. In these circumstances the haemoglobin is not able to keep all its oxygen and unloads about 30% of its oxygen very readily. What is more, if the temperature rises, or if pH drops, the haemoglobin will release even more oxygen. This can be seen in the graph on page 42 where the **oxygen dissociation** curve shifts to the *right* at higher temperatures. So, when the muscles are very active and generating much heat, the haemoglobin responds by releasing its oxygen more readily.

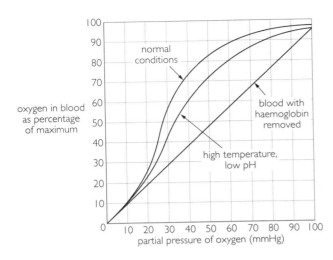

The loading and unloading of oxygen in the body normally takes place in the region where the dissociation curve is at its steepest. For this reason, the haemoglobin is very efficient at picking up oxygen in the lungs and releasing it to the respiring cells.

As well as supplying the body with oxygen, the lungs are also involved in the removal of carbon dioxide. The carbon dioxide, a by-product of respiration, is carried by the blood in three different ways: dissolved in the plasma (7%), as bicarbonate ions in the plasma (73%) and in the red blood cells (20%).

Blood which passes through respiring tissues picks up carbon dioxide and carries it to the lungs where the carbon dioxide diffuses from the blood into the alveoli. High concentrations of carbon dioxide in the blood stimulate the breathing centres of the brain to cause an increase in breathing rate. As cell respiration speeds up to produce more ATP for exercising muscles, the increased concentration of carbon dioxide causes the breathing to quicken and deepen and at the same time helps the haemoglobin carry more oxygen to the muscles. Strangely, perhaps, changes in the oxygen content of the blood have a limited effect on breathing rates.

The Absorption of Nutrients

Villi of the gut

The small intestine is around 6 metres long and much of its inner lining is folded and covered with millions of tiny projections called **villi**. These in turn have a covering of epithelial (lining) cells which themselves have many microscopic projections called microvilli. These microvilli are the product of microscopic folding of the plasma membrane of each cell. This combination of folding at three different levels of magnitude gives the small intestine a very high surface area (about 600 m²) which is needed for the absorption of digested food.

Each villus contains many blood capillaries which carry away the digested food. In the centre of each villus lies a branch of the lymph system, a tiny blind-ended tube called a **lacteal**, which carries away most of the digested fats. These fats eventually join the blood at a point near the heart where the lymph system empties its contents into the bloodstream.

Endocytosis, exocytosis, active transport and diffusion all play a part in the transfer of digested foods through the epithelial cells and into the capillaries and the lacteal. Some foods are so easily absorbed that the process starts as soon as they enter the mouth. For example, salt, glucose, alcohol and aspirin can all be absorbed in the mouth. Other foodstuffs require much assistance in crossing the intestinal barrier. For example, **vitamin B$_{12}$** needs the presence of a chemical called **intrinsic factor**. The vitamin combines with the intrinsic factor which then stimulates special cells to absorb it by endocytosis. Vitamins A, D, K and E need fats to transport them by diffusion into the blood.

The Liver

The liver and surrounding organs

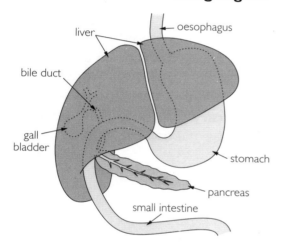

liver, oesophagus, bile duct, gall bladder, stomach, pancreas, small intestine

The liver is the largest organ of the body and, because of its great metabolic activity, produces more heat than any other organ. The functions of the liver are many, but its prime purpose is to maintain the steady concentration of substances in the blood, i.e. it is a **homeostatic** organ. For example, it is important that the concentration of sugar in the blood is kept fairly steady at around one gram per litre. Our intake of sugar varies considerably through the day. So it is the liver's function to even out these variations; to remove sugar from the blood when there is too much, and to release it into the blood when there is insufficient.

A complete list of the functions of the liver would take many pages, but a number of important functions are given below:

- Excess amino acids are converted to keto acids and ammonia in a process called **deamination**. The body cannot store proteins, so this function is essential. The ammonia is very toxic and is converted immediately to **urea** which is removed from the blood by the kidneys. The keto acids enter the respiratory pathways and can be stored as glycogen or oxidised to produce ATP.

- Some amino acids can be made from others by a process called transamination. Amino acids which can be made in this way are called 'non-essential' because they are not absolutely essential in the diet.

- Plasma proteins are manufactured by the liver, including those involved in the clotting action of the blood.

- Old red blood cells are removed by phagocytic cells. The iron is stored and transferred to the bone marrow to be used again. Some of the breakdown products of haemoglobin are excreted in the bile as **bilirubin**, giving it a green-yellow colour. The brown colour of faeces and the yellow colour of urine both derive from the breakdown products of bilirubin.

- **Bile** is produced by the liver and stored in the **gall bladder** to be released into the duodenum (first part of small intestine) from time to time to aid digestion. Bile is an alkaline fluid containing bile salts which emulsify (break up into tiny droplets) fats. Bile is an activator for the enzyme lipase but does not itself contain any enzymes.

- The fat-soluble vitamins A, D and B_{12} are stored in the liver.

- Cholesterol, a lipid, is manufactured from acetyl coenzyme A (acetyl CoA). Although excess cholesterol is thought to contribute to arterial disorders, cholesterol is in fact essential for the manufacture of steroid hormones and vitamin D, and is a major component of membranes.

- Glucose and glycogen can be interconverted under the influence of a variety of hormones (see page 45).

- Poisons (**toxins**), such as alcohol and many drugs, are detoxified and removed from the blood.

- Hormones are converted to inactive compounds in the liver, so limiting their period of activity within the body.

The blood supply to the liver is unique. Unlike any other organ in the body, it is supplied by blood from an artery and a vein. The **hepatic portal vein** brings blood from the intestines, carrying the digested food which has been absorbed. Since the intestines have used up much of the oxygen, the liver gets its own separate supply of oxygen directly from the aorta via the **hepatic artery**. After blood has been dealt with by the liver it rejoins the main circulation through the **hepatic vein**.

The Kidneys

We have two kidneys and their function is to regulate the fluid balance of the body (**osmoregulate**) and to remove nitrogenous waste such as **urea** from the blood. Approximately one thousand litres of blood pass through each kidney every day.

The process at first may seem rather unusual and inefficient. Each kidney contains around one million filtering units called **nephrons**. Around 180 litres of fluid per day are allowed to escape through tiny coiled knots of capillaries in the nephrons called **glomeruli** and then 99% of this fluid is **reabsorbed**. This requires an enormous amount of energy and the kidneys consume 10% of all the oxygen used by the body at rest. It might seem far more efficient to select poisonous molecules from the blood and only remove them. However, this would be a hazardous process for cells as they would then have to deal with high concentrations of poisonous materials. By allowing all the fluid to pass through the Bowman's capsule and then only reabsorbing harmless molecules, substances such as urea are dealt with indirectly in very dilute quantities.

Kidney nephron

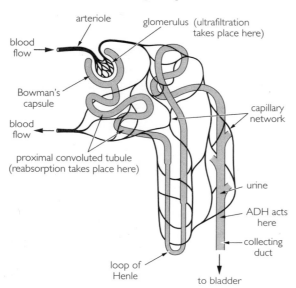

The glomeruli have tiny pores which act as micro-filters, allowing almost all the blood plasma to pass through except for cells and large plasma protein molecules. The re-absorption process takes place in a tiny coiled tubule (proximal convoluted tubule) attached to the **Bowman's capsule**. All useful substances such as glucose, vitamins, and amino acids are reabsorbed along with much of the water and salts, leaving only enough water to dissolve the urea and excess salt for excretion. This first segment of the tubule is lined with highly active cells covered in microvilli.

Further along the tiny tubule, at the loop of Henle, salts are pumped into the cortex of the kidney to increase the osmotic concentration of the tissue fluid there. When water is in short supply, the pituitary gland secretes a hormone called **anti-diuretic hormone** (**ADH**) which makes the walls of the collecting ducts more permeable to water. So water passes by osmosis from the filtrate back to the blood to make it more dilute, and, as a consequence, the urine becomes more concentrated.

Regulating Mechanisms

Cells and their enzymes are very sensitive to changes in their surroundings. For this reason the body maintains the composition of body fluids and its temperature within very narrow limits. The process is called **homeostasis** and involves sensors of several kinds and a method of control called **negative feedback**. By this process, when any factor deviates from its set point, various correction mechanisms come into play to correct the deviation. When the deviation has been corrected, these mechanisms cease to operate until another change is detected.

The Control of Blood Sugar

The concentration of sugar in the blood is held at just under one gram per litre, principally by the activities of the liver and pancreas. The hormone **glucagon** stimulates the release of glucose from glycogen stores. The hormone **insulin** has the opposite effect. Insulin and glucagon are manufactured in the **Islets of Langerhans** in the **pancreas** and it is these cells which detect changes in glucose concentrations in the blood. If glucose levels drop, glucagon is secreted, glucose levels return to normal and the stimulus for glucagon secretion disappears. If glucose levels rise, insulin is secreted, glucose levels return to normal and the stimulus for insulin secretion disappears. These are examples of **negative feedback responses**. Other hormones, such as adrenaline and thyroxine, also influence blood sugar concentrations.

The Control of Temperature

The temperature of the blood is monitored by the **hypothalamus** at the base of the brain. When body temperature rises much above 37°C, various mechanisms are brought into play to reduce temperature. Sweating results in the evaporation of water from the skin and this takes heat energy from the body. The widening of arterioles (**vasodilation**), to allow more blood to flow near the surface of the skin, helps heat to escape more easily. Conversely, shivering and **vasoconstriction** help generate and conserve heat respectively. The raising of hair traps air which is a good insulator. In the long term, the body's metabolic rate can also change to generate more, or less, heat. These mechanisms, however, are **involuntary** (not under conscious control). We can consciously help to maintain a steady temperature, for example, by seeking shelter, changing our clothes, curling up in a ball or stretching out.

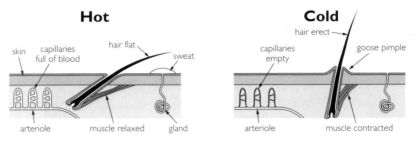

Hypothermia

Hypothermia sets in when the temperature corrective mechanisms fail to maintain body temperature and core temperature falls below 35°C. Watersports enthusiasts and mountaineers are particularly vulnerable to hypothermia, as are the very old and the very young.

Wet clothing can conduct heat from the body ten times more quickly than dry clothing, particularly when it is windy. This is because water draws heat from its surroundings as it changes from liquid to vapour. Water is also a better conductor of heat than air, so human survival time when immersed in water at 5°C can be less than one hour.

Young children are particularly at risk for a number of reasons. Being small, they have a large surface area-to-volume ratio. Since heat is generated in the 'volume' of the child and lost through the 'surface area', there is a relatively small source of heat-generating capacity and a relatively large area over which any heat generated is lost. Infants also have underdeveloped automatic temperature regulation mechanisms and have no **voluntary** responses, i.e. they cannot take conscious actions, such as putting extra clothes on, or seeking shelter, to remedy the situation.

Old people are particularly at risk because their metabolic rates decrease with age and their temperature regulation mechanisms are not so effective. For example, their vasoconstriction response tends to be weak, and their muscles are less able to generate heat energy during shivering. In addition, old people tend to be less active, they may have a poor diet and they may be reluctant or unable to warm their houses effectively in winter for financial or other reasons.

UNIT THREE
BEHAVIOUR, POPULATION AND THE ENVIRONMENT

This forty-hour Unit contains information on the nervous system, the brain, memory, human behaviour, the rise in the human population and humans' impact on the environment.

The Brain

The brain is composed of some 10^{11} **neurones** (nerve cells). Each neurone can be linked directly to hundreds or even thousands of other neurones, so the potential for almost infinite complexity is enormous. The human brain is not the largest of the animal world but, compared to human body size, it is relatively enormous. It is the computing power of the human brain which has allowed us to become conscious animals able to communicate in a sophisticated way and to engineer our environment to our liking in a way unparalleled by any other species on the planet. Surprisingly, although the human brain is the most important organ of the body, and the most significant in terms of the difference between ourselves and other mammals, relatively little is known about its functioning.

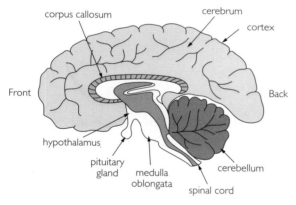

corpus callosum   cerebrum
                        cortex
Front   Back
hypothalamus
pituitary
gland   medulla   cerebellum
        oblongata
            spinal cord

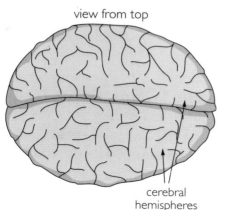

view from top

cerebral
hemispheres

The largest part of the brain is the **cerebrum**, whose **cortex** (surface) is heavily **convoluted** (folded). The folding allows an increased number of cell bodies (50 billion of them) to be located in the cortex, so maximising the potential for interconnections between neurones. In fact, the folding is so considerable that the cortex, if laid flat, would have approximately the same area as a double page of a broadsheet newspaper. From these cell bodies, a mass of **axons** (fibres) runs under the cortex to connect them to one another, to other cell masses in the brain, and to the rest of the body by way of the spinal cord. The cerebrum is divided into two halves: the two **cerebral hemispheres**, which are quite separate from one another apart from a mass of fibres, called the **corpus callosum**, which transfers information from one hemisphere to the other.

Attached to the underside of the brain at the back is the **cerebellum** which is concerned with the unconscious fine control of voluntary muscle movements and balance. The whole brain sits on the brain stem which connects the brain via the **medulla oblongata** to the spinal cord. The medulla is concerned with the unconscious co-ordination of basic functions: breathing, heart rate, digestion and **reflex actions** such as sneezing, coughing and vomiting.

Despite its great complexity, the brain is very flexible. For example, after injury, the functions of the damaged part can sometimes be taken over by unaffected parts. This flexibility of the brain is described as its **plasticity**.

The brain receives information from sensory receptors, for example, the eyes, the ears and the skin. It then deals with this information and may act on it by sending out information along motor neurones to muscles or glands. The **sensory** and **motor** areas of the brain are fairly clearly defined but many higher mental functions, such as memory, language, consciousness and emotion appear to have overlapping domains and are not nearly so easy to pinpoint.

The diagram shows the cerebral cortex and the centres of control. Not surprisingly, the size of any segment of the motor or sensory areas of the brain is directly related to the function which that area carries out. This is best illustrated by the diagram of the human shown below which has each body part drawn in proportion to the area of the cerebral cortex devoted to its motor control.

The functioning of the brain has been studied in a variety of ways:

1. by studying the effect of brain damage (split brain, stroke or other injury) on the behaviour of individuals. An individual whose corpus callosum has been cut is described as 'split-brain'. Strokes result from blocked arteries in the brain which deprive cells of oxygen. Brain cells have a high metabolic activity and can only survive for a few minutes without oxygen.
2. by using EEGs (electroencephalograms) which record the brain's electrical activity.
3. by carrying out brain scans to highlight areas of high metabolic activity in the brain.

The brain and the spinal cord together make up the **Central Nervous System** (**CNS**). This is the processing system. Nerves from the CNS radiate to all parts of the body and make up what is called the **Peripheral Nervous System** (**PNS**). The PNS simply links the CNS to the world around us and is divided into the **somatic** and **autonomic** nervous systems. The somatic system is used for all conscious activities such as speaking, writing and walking.

The autonomic nervous system controls automatically our basic body functions such as heart rate, breathing, blood pressure and perspiration (sweating). The autonomic system has two branches which work **antagonistically** (in opposition):
(1) the **sympathetic system** prepares the body for action, whereas
(2) the **parasympathetic system** encourages the conservation of bodily resources. So the sympathetic system diverts blood from our digestive system to muscles during times of stress or exercise. It also stimulates sweating, breathing, heart rate and dilation of the pupils. The parasympathetic system has the opposite effect and comes into play during times of relaxation and rest.

Neurones

Neurones (nerve cells) conduct electrical impulses from one part of the body to another. They consist of a **cell body** with numerous cytoplasmic extensions called **fibres**. The cell body contains all the usual organelles except for centrioles. This means that it cannot divide and needs to function throughout the life of its owner. Unfortunately, if neurones are damaged, regeneration is very poor. Hence paralysis after an accident is usually for life.

The fibres are of two types: **axons** and **dendrites**. Axons can be exceedingly long (up to a metre in length) and they carry impulses away from the cell body. Dendrites carry impulses towards the cell body. The axons are often covered with an insulating **myelin sheath** which increases the speed of impulse passing along the neurone. Also, the larger the diameter of the axon, the greater the speed of impulse. The myelin sheath is formed by Schwann cells which wind their membrane round the axon many times. This process is not complete at birth and this is one reason why the muscular movements of infants are less co-ordinated. Multiple sclerosis is a disorder in which the myelin sheaths of nerve fibres degenerate, so affected individuals slowly lose the ability to control their muscles.

Sensory neurone

There are basically three different types of neurone: **sensory neurones** carry information from the senses to the brain; **motor neurones** carry instructions from the brain to the muscles and glands of the body; and **association (relay) neurones** are confined to the CNS and link with one another and with sensory and motor neurones. Bundles of sensory and motor fibres linking parts of the body with the brain are called **nerves**.

Motor neurone

Neurones are linked to one another in an almost infinitely complex way but two types of neural pathway can be easily understood: **converging** and **diverging** pathways. In converging pathways, two or more fibres meet at one neurone. In the **retina** of the eye, light-sensitive cells, called **rods**, detect very weak light and pass on a weak signal towards the brain. By bringing these signals together along converging pathways, the enhanced effect is sufficiently strong to be detected by the brain. Sometimes, conflicting signals converge at a junction. In such cases one or other is dominant and stimulates or inhibits the next neurone. This can be tested with **reflex actions** such as blinking where one can try to prevent the automatic reflex by conscious will. Some people find this easy, others do not.

Diverging pathways allow one signal from the brain to be sent to many destinations simultaneously. For example, when catching a ball, all the muscles of the arm which have to be contracted or relaxed at the correct moment can do so in synchronisation.

Neurotransmitters

The point where two nerve fibres meet is called a **synapse**. At the synapse the nerve fibres do not touch. Instead there is a tiny gap (synaptic cleft) of only around 20 nm (one nm = one millionth of a millimetre) across which chemicals must diffuse to pass the impulse to the next neurone. Such chemicals are called **neurotransmitters** and there are many different kinds, the best known of which are **acetylcholine** and **noradrenaline**. Junctions between nerves and muscle fibres are very similar to synapses and are known as **neuromuscular junctions**.

Synapse

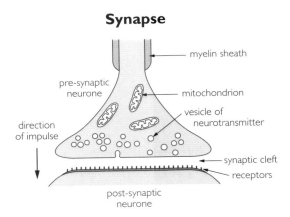

myelin sheath

pre-synaptic neurone

mitochondrion

vesicle of neurotransmitter

direction of impulse

synaptic cleft

receptors

post-synaptic neurone

When an electrical impulse reaches the synapse, it causes vesicles in the pre-synaptic axon to release a transmitter chemical which diffuses across the gap to receptors on the other side. There are many different kinds of **receptors** and they are affected in different ways by the same neurotransmitter. Incoming neurotransmitters excite some receptors and inhibit others. If insufficient neurotransmitter is produced, the stimulus is not passed on to the next neurone.

The neurotransmitters must be removed after they have acted, otherwise they would continue to have their effect. Acetylcholine, which stimulates the activity of skeletal muscle, is removed by enzyme action. Noradrenaline is reabsorbed by the pre-synaptic neurone. The continual synthesis and removal of neurotransmitters requires an enormous amount of energy and, as a consequence, neurones have a very high metabolic rate. It is for this reason the brain is so easily damaged as a result of a lack of oxygen for no more than a few minutes. It is worth noting that behaviour-altering drugs (such as nicotine, alcohol, cocaine, ecstasy and heroin, and anaesthetics used in hospitals) produce their effects by influencing synaptic transmission.

Memory

When we think of memory, we are inclined to think of our own ability to remember facts for examinations or of our memory of distant events (usually pleasant ones). However, our ability to remember is of far greater importance. For example, as you read this page, if you were unable to remember the words from the previous line then the process of reading would become meaningless. Even listening to speech requires use of memory. Some of us claim to have poor memories but we can all speak at least one language, read and write, and remember faces and facts from years gone by with little difficulty. We all have excellent memories and it is our great power to memorise language and symbols which makes us uniquely human. The brain has various areas involved in memory, one of which is the **limbic system**.

The Limbic System

Around the central core of the brain are a number of areas which together make up the limbic system. The limbic system is closely connected to the hypothalamus and seems to be involved in various aspects of memory and personality. If parts of the limbic system have to be removed by surgery, memory can be affected such that memory of distant events remains normal but memory of recent events, within the previous year, seems to have disappeared. Moreover, any subsequent happenings are not remembered. The neurones of the limbic system are rich in one particular receptor known as **NMDA** (N-methyl D-aspartic acid). In Alzheimer's disease, where memory loss is the main symptom, there appears to be a loss of the ability to make **acetylcholine** in the cells of the limbic system. These facts suggest that NMDA and acetylcholine are involved in the process of memory storage.

There are three stages in memorising of facts:

ENCODING ⟶ STORAGE ⟶ RETRIEVAL

In simple terms, when we memorise something, we turn it into a form which can be retained in our memory; we keep it there for a period of time which can be anything from a few seconds to over 100 years, and we access it from memory for use when required. The ease with which we carry out each of these functions varies enormously and depends on many other factors. For example, when we fail to remember someone's name this can be due to a failure of any one of these stages. It seems likely too that we have different forms of storage for different functions. For example, the ability to ride a bike is learned and never forgotten, but it seems that this form of memory is different from the one in which we memorise our friends' names.

What is more, our memory of words as they are spoken operates in a different way to our memory of, say, our address. When we listen or read, we use our **short term memory (STM)** to remember each word for a few seconds to enable us to understand each sentence. When we try to remember something for longer than just a few seconds we use our **long term memory (LTM)**. Short term memory has a very limited capacity. On average, the capacity seems to be around 7 items ± 2. This is called the **memory span**. It is very easy to test memory span. Simply read out unrelated lists of numbers or letters to another person and test how many they can recall correctly.

Any list longer than 8 items becomes almost impossible to remember without using other specialised techniques involving LTM.

The majority of information encoded in memory is encoded as sounds (**acoustic coding**) but we can also encode information as pictures (**visual coding**) or by using meaning (**semantic coding**). Information can be held in short term memory for about 30 seconds. For example, when you look up a telephone directory, you often remember the phone number just long enough to use it – it has been stored in STM and not transferred to LTM. You can remember the number in STM so long as it contains around seven chunks or less, e.g. 842561. What is more, most people remember a number or a series of letters like this by **rehearsing** it briefly, i.e. by saying it over in their heads a few times. This is **acoustic** coding. Visual coding and semantic coding can also be used to store memories. If we used visual coding to remember the telephone number, we would have a mental picture of the number in our mind's eye. If we used semantic coding we would place meaning round the numbers and remember the meaning rather than the numbers.

This technique enables us to remember more than seven numbers by '**chunking**' them. If the telephone number was 0141 654 1352 we might remember the first four digits as one chunk because our LTM knows that 0141 is the code for Glasgow, 654 might be remembered more easily because the digits decrease by one, and 1352 might be remembered as one item if we recognise that 13 is the number of playing cards in a suit and 52 the number of cards in a pack! In other words, we have boosted our STM storage capacity by making use of information already stored in LTM and by **elaboration** of its meaning. Similarly, it is much easier to remember items which have been **organised** into groups. If we were trying to remember cards which had been played during a card game we would find it much easier to remember them by suit (clubs, diamonds, hearts and spades) than randomly. These techniques of **chunking**, **elaboration** and **organisation** are not as silly as they seem. Memory experts make great use of such memory aids (mnemonics) to memorise long strings of numbers and names.

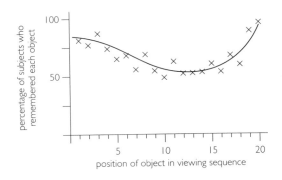

Graph to show serial position effect

percentage of subjects who remembered each object

position of object in viewing sequence

Storage

Our STM can only store information for a matter of seconds. If any extra information is added then items in STM are **displaced** (thrown out). Our STM is constantly being bombarded by new information especially as we talk, listen, watch or read, so it is not surprising that material is constantly displaced from STM. If we want to retain the information for any length of time it must be passed to LTM. The displacement of information from STM can be illustrated by a simple party game. People are shown a series of around 20 familiar objects one after the other. As soon as they have seen all the objects, they are asked to write down as many as they can remember. It is found that the objects viewed both early and late in the sequence are best remembered.

The first few objects seen are remembered better because the brain has had time to rehearse a few of them before the STM is full. Those viewed late in the sequence are relatively easy to remember because they have not been displaced from STM. Objects viewed in the middle of the sequence have been displaced by others before the brain has had time to rehearse them. This is called the **serial position effect**.

Retrieval

Although we know we have remembered something, it is not always easy to retrieve it from our memory. How often have we said 'It's on the tip of my tongue'? Sometimes it is easier to retrieve a particular fact or episode if we find ourselves in the same situation, or context, as when we encoded it. **Contextual cues** which aid recall might be smells, sounds or the sight of a particular place or person. In these situations we sometimes say that 'our memories flooded back to us'. However, we are rarely in the position of being able to return to the place where the memory was formed! If we are having difficulty remembering something we have forgotten, then the recall of the item is often aided by thinking around the situation in which it was stored. We may be trying to remember the name of someone we met at a party. Often, if we visualise the evening out, the drive to the party, the house it was in and the others whom we met, the name may be located in our memory more easily. Sometimes retrieval is aided by changing the subject entirely and leaving the brain to search for the item subconsciously.

The Development of Behaviour

All aspects of development, including behaviour, are influenced by **maturation**, **inheritance** and the **environment**. The central problem of studying behaviour is the degree to which each of these factors influences behaviour.

Stage of development	Age at which 50% reach stage
sits without support	5+ months
stands with support	6 months
crawls	8 months
walks with support	9 months
walks independently	1 year

Maturation

The development of many aspects of human behaviour is related to our age. Newborn children cannot speak, walk or grasp a ball but, as they mature, they develop the ability to do all these things. The sequence in which these developments occurs is fairly constant even between children of different races and cultures. The table opposite gives the age at which 50% of young children reach the various stages in their development.

It is very difficult to cause deviation from these stages of development, even through fairly rigorous training, and so we can conclude that the behaviours have a genetic basis and are linked very closely to age. We know, for example, the nerve fibres of very young children are not fully **myelinated** and so the nerve impulses travel more slowly in the axons.

Inheritance

It is quite clear that inheritance plays a major part in our behaviour. For example, such conditions as phenylketonuria (PKU) and Huntington's chorea (HC) are controlled by single genes and both conditions result in a deterioration of the nervous system, with consequent changes in behaviour. PKU children who remain untreated suffer severe retardation in mental activity and die before the age of 30. HC victims, although apparently perfectly normal for the first 30 years or so of their lives, gradually lose their ability to talk and control their movements, and they show a marked deterioration in memory and mental ability before they eventually die.

There are many examples of 'normal' behaviour in humans too, which clearly have a genetic basis: territorial behaviour, sexual behaviour and aggression to name but three. But can the influence of the environment override the influence of the genes? In many cases it can.

Environment

The influence of the environment on behaviour can best be studied using **monozygotic** (identical) twins and **dizygotic** (fraternal) twins, especially if they have been reared apart. If the twins are monozygotic, the effect of genes can be discounted because the twins will have an identical genotype. Any differences in behaviour must then result from differences in their environment only.

In the personality disorder schizophrenia it is known that if one twin is schizophrenic then the chances are high that the other twin will exhibit some signs of similar mental disturbance. But whether the second twin develops the full symptoms will depend on the environmental influences he or she has experienced whilst growing up. It seems that genes predispose (make more likely) individuals to certain behavioural characteristics, but whether or not they develop them further depends on their environment, especially during early development.

Intelligence

People differ in their intellectual ability, but how much of this is due to genes (nature) and how much is due to environmental differences (nurture) has been the subject of much controversy and debate. All that one can say is that both genes and the environment influence intellectual ability. Intelligence is one aspect of intellectual ability which has been studied extensively for many years. **IQ tests** used to be regarded as more important than other measures of intellect, but they do not test fine motor skills, artistic skills, social and leadership skills or the ability to memorise large quantities of factual information. They may not even have much to do with imagination and creativity. In fact, there is no general agreement as to what constitutes intelligence. But Intelligence Quotient (IQ) tests have provided a fund of information to shed light on the 'nature' versus 'nurture' debate. For example, can IQ be improved by practice or even by a change in diet? There is evidence to suggest that both are possible. Do identical twins have the same IQ even when separated at birth and reared apart? No they don't but their IQs are more closely similar to one another than to other relatives.

Intelligence Quotient (IQ) Questions

Some questions rely on your ability to sort out the meaning of words.

For example, which two words in brackets are connected in the same way as those underlined?

<u>cruel – kind</u> (angry, generous, unhappy, timid, selfish, shy)

<u>mouse – mammal</u> (thought, sugar, taste, emotion, water, fear)

Some questions rely on your ability to spot sequences.

For example, what are **x** and **y** in the following sequences of numbers:

1,2,4,7,11,x   3,5,4,6,y,7,6,8,7   Now try this question:

If 2 and 4 with 3 = 5; 4 and 3 with 2 = 10, what does 2 and 3 with 4 equal?

Some are more subtle and rely on more devious powers of deduction.

For example, what is the next letter in each of the following sequences:

o,t,t,f,f,s,s,e,n, ...   m,t,w,t,f,s,s, ...   c,f,i,l,o,r, ...

Infant Attachment

Human children spend a much longer time 'growing up' than do other young mammals. The evolution of sophisticated communication skills is unparalleled in other species. This has resulted in the need for an extremely long period of time for learning and consequent **dependency**. From the moment they are born, children are highly dependent on their parents and this dependency requires strong bonds to be forged between the infant and its parents at a very early stage. Crying, smiling and suckling all provoke powerful responses in the parents. This attachment can also be seen in many other higher animals and has obvious survival advantages, as the young are more likely to obtain food, warmth, comfort and protection from their parents as they grow. But in humans the extra time required is for the enormous amount of learning which takes place.

One might guess that the most important of these early needs would be food, but experiments with monkeys often show that young monkeys rate warmth and body contact higher than food. Infant monkeys separated from their mothers at birth will choose a warm, cloth model which provides no food, in preference to a cold wire model which provides them with milk. If the model rocks, the young monkey will be even more attached to it! When all these features are tested separately, it appears that contact with a soft cuddly model is preferred by infant monkeys.

However, even a soft warm model mother which rocks and provides food is not enough to ensure infant monkeys develop satisfactory 'normal' behaviour. Young monkeys reared in such an artificial situation have difficulty coping with other monkeys and show inappropriate sexual behaviour. What is more, females raised in this way turn out to be poor mothers, tending to neglect, or even harm, their infants.

It would be dangerous to assume that everything described above applies to humans, but equally it would be foolish to assume that the general principles are not the same. It is certainly true to say that the links a child makes with its mother and, to a lesser extent, its father, in its early years, are vital in establishing its ability to cope with adult life, to establish sound relationships with others, and to be an effective parent.

There appears to be a relationship between the pattern of early attachment in humans and the way an infant copes with new experiences in subsequent years. Children who have a secure upbringing very often approach problems with enthusiasm and persistence. When they encounter difficulties they seldom give up and do not show many signs of frustration. Those who have experienced insecure attachment as infants become easily frustrated, give up readily and do not seek or welcome adult assistance. Children observed as securely attached early in life tend to be social leaders and are sought out by other children. Teachers rate them as self-sufficient and eager to learn. Insecure children tend to be socially less able and reluctant to participate in group activities. Teachers see them as less curious and less forceful in pursuing their goals, and these differences are not related to intelligence. These trends are general, of course, and there are many individual differences between children with otherwise similar backgrounds and experiences.

Communication

Communication is the transfer of information from one individual to another. All animals can communicate, and some do so in sophisticated ways, but none can approach the amazing complexity of human communication skills. It seems certain that our **innate** ability to learn language is unique to our species. Many humans cannot read and many cannot carry out simple arithmetic, but almost all can easily communicate by use of complex spoken language.

Infants, before they are able to speak, communicate most effectively in a **non-verbal** way with their mothers. The infant smile is a particularly powerful signal to its mother. However, non-verbal communication takes place between adults too, often without us being aware of it. Subtle movements of the body and the face give much information to others about our mood, intentions or our deceptions, even when we try to convey the opposite information through language! In fact, when verbal signals contradict non-verbal signals, the effect can be most powerful.

Interestingly, some non-verbal signals vary in the information they convey from one culture to another. In Russia the slow hand clap is complimentary. In Italy the subtle difference between the two-fingered gesture (V-sign), palm out or in, is not recognised, so Italians usually respond positively to the V-sign whether given palm out or in! The thumbs up signal we use to attempt to gain a lift from a passing car is translated to 'sit on this' by Sardinian drivers!

All parts of our body are used in non-verbal communication – the way we stand, the use of our hands, even the clothes we wear signal what we are. In schools where uniform is worn it is still quite possible for students to convey information about themselves by the manner in which they wear their uniform.

We do not look at only one part of a person's face when we look at a face for several minutes. Instead, we look around the person's face, spending different amounts of time on different parts, with most time spent on the eyes and mouth. The example above shows what parts one person looked at over several minutes of looking at this face.

However, when we communicate with others we invariably find their faces most interesting. Most of human non-verbal signalling is carried out using the multitude of facial muscles which we have acquired through millions of years of evolution. Even the whites of our eyes have evolved so that eye movements are detected more easily by others. The human face is far more mobile than that of any other primate. In particular, the eyes and the mouth convey a huge variety of non-verbal information. When we look at someone who is speaking to us, our eyes constantly check their eyes and mouth more than any other part of their face.

Language

A great deal of our information about brain mechanisms for language comes from observations of patients suffering from brain damage. The damage may be due to tumours, penetrative head wounds, or the rupture of blood vessels. From these studies we have been able to find out which areas of the brain deal with the various aspects of speech and communication in general. For example, we know that the left side of our brain governs the ability to express ourselves in complex language, and the right hemisphere can only comprehend very simple language.

All languages have a limited number of speech sounds. In English we distinguish around 40 different sounds but an average adult will understand around 40 000 words. In addition, rules for combining words make it possible for humans to understand billions of sentences, and by the age of five we have accomplished most of the task! Children can speak a few isolated words at the age of 1; by the age of 2 they can speak in two- or three-word sentences; by the age of 3 the child's grammar is becoming more complex, and by the age of 4 the sentences are quite sophisticated. By the age of 6 most children have a vocabulary in excess of 15 000 words.

Can other primates communicate in the same way? Attempts have been made to try to get chimpanzees or gorillas to learn to speak. Speaking is beyond them because of their lack of fine motor control of voice box, tongues and lips. However, it is possible to teach them to communicate by sign language of one kind or another. There are cases of apes being taught a few hundred signs or signals, but it appears they cannot construct sentences with these symbols. Moreover, humans can place the words 'chimpanzee', 'chased', 'the' and 'man' into a sentence: *The chimpanzee chased the man.* whereas apes cannot. What is more, humans can switch the same words around to convey a different meaning: *The man chased the chimpanzee.*

How do children learn to speak? Do they simply imitate adults, or is there more to it than this? **Imitation** certainly plays a part in learning. That is why French children speak French and why English children speak English. Moreover, parents will point to an item such as a spoon and repeat the name of the item to the child so that it can attempt to imitate the sound. There is much evidence, however, to show that children learn to produce sentences by quite a different method. Young children are constantly producing meaningful sentences which they quite clearly have never heard before, e.g. 'All gone milk!' In addition they generate words which do not exist: 'Daddy taked me to swim pool.' This shows that they are applying general grammatical principles rather than copying what adults say. In reality, it is impossible for children to copy everything they hear. It is estimated that the number of twenty-word sentences that we can understand is of the order of 10^{20}. To learn such sentences by repetition would take a child many billions of years if the sentences were simply spoken once each at normal rate! There is much other evidence (from comparison of different languages and from studies of children brought up in an environment where no one speaks) to show that the learning of a language is a complex process, much of which is innate. Moreover, the processes are quite unique to humans. This powerful tool which we have at our disposal has resulted in our rapid cultural and scientific evolution.

The Effect of Experience

If a hungry cat is placed in a cage with a simple latch and a piece of fish is left outside, it may eventually learn to lift the latch to get to the fish. If this situation is repeated often enough, the cat will learn to respond very quickly. It might appear that the cat has solved the problem in the same way as a human might look around, assess the situation, and make intelligent guesses as to what to do. But cats are not as smart as humans and, even if the cat's paw is put on the latch by the experimenter a few times, the cat seems unable to make any kind of logical deduction. It is unable to **imitate** our action. Instead, the cat hits upon the solution by accident, and it is only after it has solved the problem in this way a number of times that it starts to associate its action with the reward. This kind of behaviour, performed by many animals, including ourselves sometimes, is called **trial and error** behaviour. The type of reward, the frequency with which the reward is given, and the time interval between response and reward, all have an effect on the behaviour of the cat – and us for that matter! Instant rewards **reinforce** the behaviour better than rewards which follow some time later. Strangely, intermittent rewards reinforce activity more than regular rewards. Perhaps this is why fruit machines are so addictive! If rewards cease, then the behaviour undergoes **extinction**, i.e. it disappears.

If the cat learns that the lever will only open the door when a light comes on in its cage, then it has learned to **discriminate**. If the cat is taught to perform an increasingly complex series of tricks by an experimenter, this process is called **shaping**. Each action that the cat performs which approximates to the wishes of the trainer is rewarded, and progress is made towards the ultimate goal of the trainer.

Animals can also be taught to behave in particular ways by punishing them for the wrong action, rather than rewarding them for the right action. This form of reinforcement can be less effective, because the animal only learns what *not* to do, rather than what to do. But in some circumstances, pain can be a very powerful motivational force.

Much can be learned about animals' responses to rewards and punishment, and such knowledge can be related to human behaviour. Young children respond better to instant intermittent rewards for good behaviour rather than delayed punishments for bad behaviour. This is not an argument for abandoning punishment regimes however! It is not hard to imagine situations where a punishment rather than a reward fits the action. In truth, many reward children or animals at the wrong time and promote bad behaviour as a consequence. We must beware of giving in to a dog's or a child's demands with some form of reward, because we are producing quite the opposite behaviour from that which we intend! An example of this might be the mother who gives her child a sweet to stop her crying at the supermarket checkout. The child quickly learns that crying at this point in the shopping trip provides a tasty reward. If the child is ignored or rebuked at this point, then the behaviour will probably become extinct (for ever one hopes!). But many parents still take the soft option and everyone suffers – not to mention the damage caused to the child's teeth! Similarly, if a child cries in bed at night until someone comes to comfort it, then this behaviour is reinforced. However, it is not wise to ignore crying children, particularly if they are very young. There may be a very good reason why a particular child is crying and our attention is needed. What is more our genes have programmed us to be very conscious of such powerful signals, and not without good reason.

If a child is bitten by a dog, or scratched by a cat, it may cry and it would be wrong to ignore its cries. Following this incident, the child may then be frightened by all dogs or cats. If this happens we say that the child's behaviour has become **generalised**. The opposite type of behaviour is called **discrimination**, where a child might be taught that the family pet is quite safe to handle but the Rottweiler down the road is not!

Humans are particularly good at **imitation**. In fact, a great deal of human behaviour is learned by observing and imitating the behaviour of others. This is often a preferred means of learning and is used in training, e.g. when learning to ride a bike or play a piano. Here, **practice** plays a key role in establishing the **motor skills** required. By continued imitation and practice, neural pathways are established in the brain which can remain in place for the rest of one's life.

Group Behaviour

Humans are social animals with very complex cultural rules and a great deal of human behaviour results from influence by others. Our parents have the most profound influence on our behaviour, followed by brothers and sisters (siblings), friends of the same age (peers) and teachers. We are also influenced by people we admire (e.g. by film stars and sports stars) and we are influenced by what people do. If we were not, then advertisers would cease to advertise.

We may change our behaviour (e.g. stop smoking) because of something we have heard, seen or read, which convinces us that it is indeed worth changing, for good reasons which we believe to be true (better health in the case of smoking). This more permanent alteration of behaviour is known as **internalisation**.

We may change our behaviour, not because we believe in any intrinsic value in the behaviour we wish to copy but because we know someone whom we admire or respect behaves in this way and we wish to be like them. This alteration of behaviour is known as **identification**. A boy or girl might take up a sport played by a person he or she admires. Trends in fashion are often brought about by our wish to be like people we admire.

We may perform a task more effectively in the presence of others than we might do on our own. This alteration of behaviour is know as **social facilitation**. It is worth noting, however, that the complexity of the task influences the outcome of this form of behaviour. In general, people will run faster if watched by spectators. They will also tend to run faster if accompanied by another runner. However, although people will perform more arithmetical problems in a given time when others compete against them, they will produce more errors than they would if they had completed the activity on their own! We might simply say 'they were nervous!'

We behave differently when we are with a crowd of others. In such situations we can find our behaviour quite different from what we would consider as being normally acceptable. This alteration of behaviour is known as **deindividuation**. Deindividuation can result in unpleasant and disturbing human behaviour, for example, in the case of aggressive behaviour of football supporters during or after a match. In these situations, the behaviour of others is copied to ensure the acceptance of the group. Bad behaviour by the individual is thought to be admired by others and be less likely to be discovered or punished.

The Population Explosion

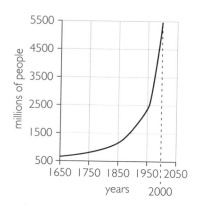

It is probably true to say that most of the major problems we face today are a direct result of the fact that there are increasing numbers of humans in most parts of the world.

The growth of the human population is not a new phenomenon. It has been going on since the emergence of our species less than a million years ago. For most of our evolutionary history, our population has shown slow but consistent growth. However the rate of growth has become explosive in recent times and shows little evidence of slowing down.

Because populations in some developed countries, such as our own, are relatively stable and because the greatest population increase is taking place in the poorer countries, it would be easy for us to say that the blame lies elsewhere. But it must be remembered that the population densities of some of the most advanced countries of the world are very high and their consumption of resources is much higher than that of the poorer countries. The United States contains only 6% of the world population but consumes 30% of the world's energy resources. India supports almost 20% of the world population but only uses 2% of its energy. Hong Kong has a population density of 300 000 per km^2. Europe has a population density of 100 per km^2. But Africa has a population density of only 15 per km^2.

So what are the reasons for the increase in the human population?

The first small populations of human ancestors appeared in Africa more than four million years ago. Since then, humans have spread out to inhabit practically every land area on Earth. Today there are over 5 500 000 000 people inhabiting the Earth and the population is increasing by about 2% per annum. This may not sound very much, but if this rate of increase of three persons per second continues for the next 500 years there will only be enough room for one person per metre^2 of the entire land surface of the Earth!

Before the invention of agriculture our ancestors depended on hunting and gathering as a means of obtaining food. Consequently, their population was regulated by what they were able to obtain from their environment and our estimates of their population size are based on our knowledge of the optimum numbers that can be sustained by this way of life, i.e. by knowing the **carrying capacity** of the Earth. By comparison with the populations of hunter/gatherer societies of humans living in the jungles or in the deserts today, it seems that around five to ten million humans inhabited the Earth before the agricultural revolution. This is around the same number of people who live in Scotland today! Primitive human populations, like all animal populations, were held in check by predation, availability of food, disease and climatic factors such as cold and drought.

Moreover, nature operated its own form of contraception. Young children were carried by their mothers for many years and could suckle on demand almost to the age of five. Suckling stimulates the production of hormones which inhibit ovulation, so women were unable to become pregnant for a number of years after the birth of each child. Today, many babies are bottle fed, and those who are breast fed cannot always obtain a drink when they wish, so this natural form of contraception has virtually disappeared from all human societies.

It is likely that the human population only increased very gradually for many thousands of years. However, about ten thousand years ago the population started to increase more dramatically as humans learned to cultivate crops and domesticate animals – the beginnings of **agriculture**. By about 2000 years ago, the world population had reached around 200 million, and it doubled in the next thousand years. Since then the rate of increase has been even more rapid and, at the present rate of increase, the world population is doubling not every thousand years but every forty years.

The **Agricultural Revolution**, which started ten thousand years ago, resulted in a huge increase in food production which continues today to outstrip population growth in most parts of the world. For this reason, many argue, we should not be concerned about population growth.

The **Industrial Revolution**, which started in the 1800s, brought with it great improvements in both agriculture and medicine: it improved **food supply** and improved **health care** and resulted in a massive decline in the death rate, particularly of young children. Unfortunately, then and now, birth rates in many parts of the world have changed little since humans first appeared on Earth.

It is important to realise that most animal and plant populations do not show long term growth of the kind we see in humans. Instead, animal and plant populations remain relatively stable for long periods of time, fluctuating only within narrow limits. Major changes in populations normally only take place in response to marked changes in the environment's capacity to support a particular species – its **carrying capacity**. Furthermore, most populations are stable, in the sense that if they are affected by some environmental event, a fire or a particularly cold winter for example, they quickly return to their previous size when conditions return to normal. It follows, therefore, that in such stable populations, **birth rate** matches **death rate**, or some **emigration** or **immigration** occurs to balance the books. In a growing population, such as our own, birth rate exceeds death rate. If we consider the human population of the world as a whole, then emigration and immigration have no effect on the total population unless we board spaceships and leave the Earth!

So what solutions are on offer? In Europe and North America, the human populations are relatively stable, and some are even in decline. There has been a **demographic** (relating to population) change from high birth rate and high death rate to low birth rate and low death rate. The lower death rate has been brought about by improvements in medicine and sanitation; the lower birth rate has resulted from the availability of increasingly effective forms of contraception and abortion – whether we approve or not. The decision to limit family size has been mainly due to simple economics. In primitive agricultural societies, children were regarded as an economic bonus (more farm workers) and, since many died of disease, it was important for a family to produce as many babies as possible. However, as societies became industrialised and medicines improved, children were more likely to survive and be an economic cost (rather than an economic bonus) to the family. Today in Western society, children could be regarded as very expensive luxuries! At present, the cost of raising children is, by and large, the only factor limiting population growth in developed countries. In underdeveloped countries population growth is still considerable, with a few exceptions. For example, in China and India, where the human populations are enormous, the governments have introduced a number of laws and financial incentives in an attempt to limit population growth. So the future looks uncertain, but one thing is absolutely certain, if the human species does not reduce its birth rate voluntarily, or by compulsion, then early death must follow for many.

Feeding the Human Population

One of the most important factors which has permitted the human population to grow rapidly, particularly over the last two hundred years, is the revolution in the production and transportation of food. Ten thousand years ago, humans started to tame wild animals and breed them for meat and milk and easy handling. In addition, they started to cultivate wild plants in 'fields' from which other plants (and animals) were excluded – as **monocultures**. Natural vegetation was removed to allow for this new found activity – farming. As a consequence, there was no need for a nomadic life and humans could settle in communities – villages, towns, cities and now, 'mega-cities' containing millions of humans. To feed this growing human population:

- forests continue to be cut down to make way for fields
- deserts are made fertile with irrigation schemes by building wells, dams and canals
- mountain sides are terraced for the growing of crops such as rice
- areas of land are placed under glass to protect them from weather extremes and to raise temperatures artificially
- the sea is kept at bay with dikes, and the salt washed from the mud to enable crops to be grown
- crop rotation is employed to reduce uptake of the same minerals from the soil from year to year and to keep one step ahead of invading pests
- artificial fertilisers (nitrates and phosphates) are used to raise the productivity of poor soil
- pesticides are constantly improved to stay one step ahead of mutating insects, fungi, bacteria and viruses which damage our crops and kill our animals
- herbicides are used to kill unwanted plants
- hormones are developed to increase growth rates
- antibiotics are used to kill animal pathogens
- large numbers of animals, such as chickens and pigs, are kept under one roof and reared intensively, often with scant regard for their wellbeing
- new machinery is invented to speed up planting and harvesting of crops
- refrigeration techniques are used in the transport and storage of produce...

... there is no end to human ingenuity.

Selective breeding is one area where enormous advances have been made recently in the development of new strains of plants and animals. For thousands of years, humans have selected the best seed and the best young animals and bred them to produce desirable traits such as disease resistance, fast growth, good yield, flavour, colour and so on. However, recently the pace of artificial evolution has quickened with the development of **genetic engineering**. Genes can now be selected for valuable traits and transferred from one organism to another. Breeding can take place in the test tube or petri dish and **clones** (genetic copies) of animals and plants can be produced very quickly and relatively easily.

Despite all the advances, however, the provision of food for the world population is very uneven and many millions live at starvation level while others, in more technologically advanced societies, produce a superabundance of high quality food. In these countries, farmers and fishermen are paid *not* to produce more food, and vast stocks are stored in huge warehouses. The transport of surplus food would be relatively easy, but the costs are such that the wealthy nations, rightly or wrongly, are reluctant to be so generous. What is more, the provision of food on its own is regarded by some as unwise as it offers only temporary respite and simply encourages further population growth in areas already unable to sustain their current populations.

Tragically, one thing is certain: despite all the technological advances and the overproduction of food in certain areas of the world, many millions are yet to die of **malnutrition** and **starvation** before a global solution is achieved.

Water Supply and the Environment

The worldwide need for a supply of fresh water is increasing for two simple reasons: (1) there are more humans on the planet, and (2) each human is using more water per day, directly or indirectly. Not surprisingly perhaps, the developed countries are those with the greatest demands. In the poorest countries, the consumption per person is less than 10 litres per day, whereas in the UK the figure is nearer 500 litres per day. The reason for this difference is that the UK, like other developed countries, uses a large proportion of its fresh water supplies to cool power stations, to irrigate fields and to supply industry. Also, each household uses much more water per day. For example, in the past 30 years the household consumption of water per person in the UK has almost doubled as more water is used to wash people, clothes, dishes and cars and water gardens.

As the world becomes increasingly industrialised, the global requirement for supplies of fresh water will continue to rise. However, there are two major problems which result from this increasing industrialisation:

1. As forests are removed, many areas, which once had an adequate rainfall, are now experiencing drought. Over 50% of the rain landing on a tropical forest evaporates back to the atmosphere to form clouds and later, rain. If the forest is removed, the water, instead of evaporating, runs down the hillsides to the sea. What is more, the foliage and roots of the trees protect and bind the soil. So, when the trees are removed, the soil is exposed to the drying effect of the sun and the full force of rain. The soil is blown away by the wind or washed away by the rain, leaving barren **marginal land** which is unable to support crops for more than a year or two. Dams and irrigation channels, built to preserve and redistribute water, then become clogged by the silt and are useless within a few decades of their construction.

2. As the Industrial Revolution spreads across the world, there is a consequent increase in the pollution of fresh water and sea water. Poisonous gases from power stations and car exhausts dissolve in rain making it both acidic and toxic. The rivers and oceans of the world are still regarded by most people as convenient dumping grounds for untreated sewage and industrial waste. Pesticides and fertilisers are washed from agricultural land to pollute rivers and lochs. Poisons enter the **food chains** and can be passed along from animal to animal, becoming increasingly concentrated at each link in the chain. At the end of the food chain the poisons are so concentrated that the animals are severely affected and may become infertile or die. Many of our birds of prey have been affected this way and, more importantly, we humans too are at the end of many food chains.

How DDT builds up in food chains

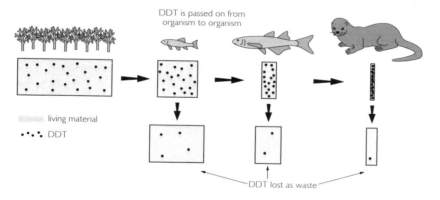

It has been predicted that within the next twenty years water will have the same value as crude oil. Nations, in their desperation to secure dwindling supplies, will resort to conflict. For example, in the Middle East, Turkey controls the headwaters of both the Euphrates and the Tigris and has built over 30 dams. The Euphrates flows through Syria which plans to develop huge irrigation schemes for its agriculture. However, both rivers then flow through Iraq which also has a right to steady supplies. Similarly, the Nile flows through Uganda, Ethiopia and Sudan before reaching Egypt. Mexico City, with a population of 20 million, is sinking in places at a rate of 4 cm per year as water is drawn from underground to supply the population. As a consequence, supply pipes are fracturing and vital water supplies are lost. Disastrous irrigation schemes to feed the Soviet cotton plantations have led to the Aral Sea in central Asia all but drying up. As it shrinks, poisonous salts in the water have become so concentrated that all the fish have died and the drinking-water is now toxic. Winds pick up the dried salts and dust and blow them over a wide area, contaminating crops and causing severe health problems to the population.

Disease and the Human Population

Without doubt, disease has limited the expansion of the human population much more than any other factor such as war or famine. The plague is a bacterial disease spread from rats to humans by fleas and from human to human by coughing. It is estimated that a third of the population of Europe was wiped out by an epidemic of the Black Death, as it was called, which started in 1347. For thousands of years, although disease killed a large proportion of the human population, this was compensated for by a high birth rate. Now, if birth rates remain high, but the treatment of disease improves, the potential for population increase is enormous. Certainly, the huge expansion in the human population over the past two hundred years has resulted from our ability to control and treat disease effectively. This has been as a result of better **sanitation**, **hygiene**, **nutrition** and **health care**.

Two hundred years ago the **mortality rate** for infants in London was around 30% (300 per 1000). Today the rate is nearer 0·7% (7 per 1000). Then, life in the city was extremely unpleasant. Houses did not have bathrooms; the unpaved streets smelt disgusting as people walked in a mixture of mud and excrement; drinking-water from the Thames was heavily contaminated with sewage and people regarded it as unhealthy to wash. This was not surprising, because it was through water-borne diseases, such as cholera, dysentery and typhoid, that most died. Cures for diseases were often more dangerous than the disease itself. For example, venereal diseases were treated with mercury compounds. Mercury poisons the nervous system causing mental disorders and violent twitching. Surgery was extremely primitive. There were no anaesthetics and no knowledge of infection and how it occurred. If you were seriously injured you rarely recovered. Patients were knocked unconscious or given large quantities of alcohol before operations. Diet was very poor. Food became quickly contaminated as there was no refrigeration. Parasitic contamination of meat was common and nearly everyone was infected by intestinal worms. Tapeworms cause internal bleeding and can sometimes block the entire intestinal system. Food was usually washed down with alcohol and many suffered from kidney stones and gout as a consequence. Not surprisingly, life expectancy was very short.

However, in the nineteenth century there was a great advance in the understanding and treatment of disease. Smallpox, a viral disease, killed over a million people each year in Europe around the beginning of the nineteenth century. At that time, a British doctor, Edward Jenner, heard a milkmaid claim she could not catch smallpox because she had suffered cowpox. He then tested her hypothesis on a young child whom he infected first with cowpox, a mild disease, and then, at a later date, tried to infect with smallpox. The child remained healthy. The concept of **vaccination** ('vacca' is Latin for 'cow') quickly gained acceptance, and today smallpox has been completely eradicated by a worldwide programme of vaccination organised by the **World Health Organisation** (**WHO**). Many other diseases, such as polio, measles, diphtheria, whooping cough and tuberculosis, are now rare in developed countries as a result of vaccination programmes.

Later in the nineteenth century the French chemist Louis Pasteur discovered that the souring of wine and milk was brought about by microbes and went on to suggest that microbes were the cause of disease rather than being the result of disease. He also developed vaccines for two deadly diseases, rabies and anthrax. Lister, an English surgeon, reading of Pasteur's discoveries, was the first to carry out operations in **antiseptic** (clean) surroundings, towards the end of the century. At the same time Koch, a German scientist, identified many **pathogenic** (disease-causing) bacteria under microscopes and developed techniques for growing them on agar jelly.

Population pyramids

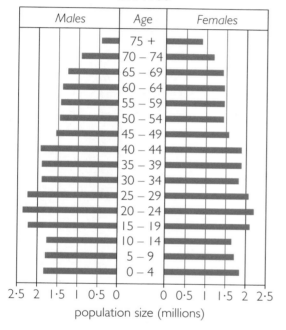

Despite these great advances, many millions of people in underdeveloped countries still die of diseases which have, by and large, disappeared from developed countries. Water-borne diseases, such as cholera, dysentery and typhoid, are a particular problem resulting from poor sanitation and contamination of drinking water. Only through improved sewage disposal, the provision of clean drinking water, improved diet and widespread vaccination will these diseases eventually be brought under control.

The diagrams show two population pyramids, one for a developed country and the other for an underdeveloped country. In the developed countries (e.g. UK), birth rates and death rates are similar and infant mortality is low, so almost all the children born go on to survive to old age. In the underdeveloped countries (e.g. Bangladesh), many die young and the pyramid takes on a much more angular shape. The study of population statistics and the distribution of age groups within populations is called **demography**.

The Nitrogen Cycle

Nitrogen is an essential element present in all nucleotides and amino acids. Without it, life as we know it would not exist. Almost 80% of the atmosphere is nitrogen – around 10^{15} tons of it – but, perhaps surprisingly, only a few organisms can make use of this massive supply of nitrogen directly. These organisms are mostly **bacteria** living freely in the soil or inside the roots of some plants, e.g. clover. The bacteria take in nitrogen from the air and convert it to nitrates and ammonia which can then be absorbed by plants through their roots. Animals obtain this combined nitrogen by eating either plants, or animals which have eaten plants. The nitrogen removed from the soil, however, is soon returned. Urine and faeces, and the bodies of dead plants and animals, are all degraded in the soil by other bacteria and fungi, which decompose the proteins and nucleic acids present in the organic material. The nitrogen in these compounds is then converted back to nitrates which can be absorbed by the plants again. In this way, nitrogen is recycled in an economical and efficient way by nature. Some nitrogen is lost inevitably, dissolved in rainwater and carried off to the sea or converted back to nitrogen gas by denitrifying bacteria (also found free living in the soil). In essence, the system has worked this way successfully for millions of years.

The Nitrogen Cycle

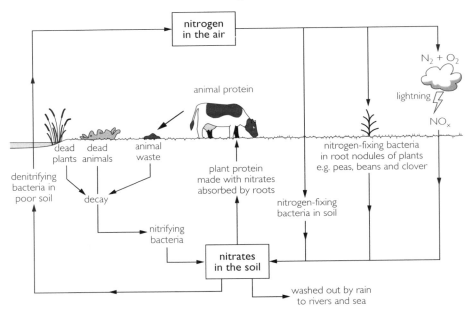

Farmers soon discovered that the addition of extra nitrogen to the soil enhanced the growth of plants. At first, organic nitrogen in the form of animal manure was added to the fields but soon inorganic nitrogen in the form of artificial fertiliser was manufactured using large quantities of energy from fossil fuels. It was inevitable that such a massive addition of nitrogen to the soils of the world would disturb **ecosystems** and poison waterways and even the seas. Fertilisers and animal waste from intensive farming eventually find their way into rivers, lochs and the sea. There they have the same effect as they do on the land – they stimulate the growth of plants. Algae grow particularly well and the **algal blooms**, as they are called, have the effect, perhaps surprisingly, of reducing the oxygen content of the water in which they grow. Moreover, some algae are themselves toxic. The increase in organic material in the water is so great that many bacteria are able to feed off the dead remains of this huge **biomass**. In so doing they remove oxygen from the water to such an extent that other animals die, particularly at night, when photosynthesis does not contribute any oxygen to the ecosystem. This is known as eutrophication, and is now widespread throughout the world, affecting our lives in many ways. Fish stocks are depleted, swimming becomes dangerous for humans, and the excess nitrogen causes havoc to fragile ecosystems. For example, the Great Barrier Reef is threatened with destruction by starfish which thrive on the coral-forming polyps. It is thought that the huge increase in the number of starfish has been brought about by nutrient pollution from the mainland of Australia. Ammonia in the air, dissolved in rainwater, is now enriching barren moorland and affecting the growth of plants adapted over millions of years to a low nutrient environment. Of all these though, the factor which is most likely to encourage remedial action by humans, is the suspicion that dissolved nitrates in drinking water can have harmful effects on growing children. What these effects are is not certain yet, but there is some evidence that excess nitrogen may be carcinogenic (cancer-causing) and may also cause growth deformities.

The Carbon Cycle

All living things consist mainly of the elements carbon, hydrogen, oxygen and nitrogen, present in virtually the same proportions in all organisms from bacteria to humans.

Carbon is as essential for life as water. Because of its ability to share up to four electrons (valency) with other elements, it can combine with them in an almost limitless way. So carbon atoms are found in almost all the compounds made by living things, and such compounds are said to be 'organic'.

Element	Percentage composition by mass (excluding water)	
	bacterium	human
carbon	51	54
oxygen	20	19
hydrogen	10	8
nitrogen	10	9

Carbon atoms get into the bodies of living things through the process of photosynthesis. All green plants remove carbon dioxide (CO_2) from the air, and water (H_2O) from the soil to manufacture glucose ($C_6H_{12}O_6$) using light energy obtained from the sun. The glucose is then converted, by the addition of various minerals such as nitrogen, phosphorous, potassium and sulphur, to the great variety of organic compounds needed by the plant.

UNIT THREE
Behaviour, Population and the Environment

The Carbon Cycle

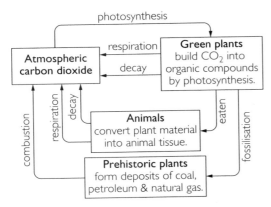

Carbon emissions in 1990

Country	Carbon emissions (in thousands of tonnes)	Carbon emissions (kg per head of the population)
USA	924 000	3500
Japan	215 000	1700
Britain	197 000	3400
China	120 000	60
India	58 000	100
Uganda	500	20

Animals then eat plants directly (herbivores) or indirectly (carnivores) to obtain their own supply of carbon compounds. In so doing they obtain much chemical energy. This energy is released during the process of respiration where the carbon is returned to the air as carbon dioxide. So carbon is continually recycled in a natural way through **photosynthesis** and **respiration**. Bacteria and fungi in the soil digest the dead organic matter and return the carbon to the air through respiration.

However, humans have disturbed this natural cycle by burning carbon compounds and releasing extra carbon dioxide into the air. Plants and animals living millions of years ago in swamps and peat lands where there was little decay have, over the centuries, turned to coal, oil and gas. These **fossil fuels** are rich in carbon compounds, e.g. **methane** (CH_4) which we burn to drive our industrial processes and warm our houses. Also, by burning the forests of the world we convert solid carbon, in the form of wood, to gaseous carbon in the form of carbon dioxide. Not surprisingly, this activity appears to be having an impact on the Earth's climate. The carbon dioxide, and other gases such as methane, which is produced as a result of the metabolism of bacteria in the soil and in the guts of animals, appear to be having a **greenhouse effect** on the planet. The heat of the sun is not allowed to escape from the Earth quite so easily as before and **global warming** is occurring. This phenomenon is almost certainly the cause of the weather extremes we are now experiencing round the world – abnormal temperatures, storms, floods and droughts. If we continue to add greenhouse gases to the atmosphere it seems very likely that there are going to be even greater changes to the climates of the world; the increase in temperatures will cause icecaps to melt and seawater to expand and rise to flood areas where at present millions of humans are living.

So, as the number of humans inhabiting the world increases by the second, the destruction of the natural habitats affects the air we breathe, the food we eat, the water we drink and the climate in which we live. The precious soil is washed away and many species of animals and plants which might have been of great medicinal or food value are driven to extinction. Never before in the history of the Earth have so many wonderfully varied and beautiful ecosystems been damaged or destroyed so quickly or so effectively. The most destructive animal ever to have lived on the planet is conducting a giant global experiment which it is reluctant to halt and which risks its own very existence.

Answers to questions on page 55

Cruel and kind are opposites, as are **generous** and **selfish**.

A mouse is a mammal. **Fear** is an **emotion**.

16 (difference in numbers increasing by 1)

5 (add two, take away one)

$(2 \times 4) - 3 = 5$, and $(4 \times 3) - 2 = 10$, so $(2 \times 3) - 4 = $ **2**

t (first letters of numbers)

m (first letters of days of week)

u (every third letter of alphabet)